SpringerBriefs in Applied Sciences and Technology

PoliMI SpringerBriefs

More information about this subseries at http://www.springer.com/series/11159
http://www.polimi.it

Daria Casciani

The Human and Social Dimension of Urban Lightscapes

POLITECNICO
MILANO 1863

Springer

Daria Casciani
Department of Design
Politecnico di Milano
Milan, Italy

ISSN 2191-530X ISSN 2191-5318 (electronic)
SpringerBriefs in Applied Sciences and Technology
ISSN 2282-2577 ISSN 2282-2585 (electronic)
PoliMI SpringerBriefs
ISBN 978-3-030-57164-1 ISBN 978-3-030-57165-8 (eBook)
https://doi.org/10.1007/978-3-030-57165-8

This Springer imprint is published by the registered company Springer Nature Switzerland AG
The registered company address is: Gewerbestrasse 11, 6330 Cham, Switzerland

To my Masters and to my beloved ones.
To Achille.

Preface

In recent years, the human and social dimension of urban lightscape has become a new theme of interest of both research and design practice in the lighting domain. The focus of this book is to explore new criteria and characteristics of the human- and social-scale lighting by considering human psychology in the design of the lighting of modern nocturnal cities. The book identifies a new area of lighting design research and practice that focus on the nocturnal urban experience of people in terms of emotional and cognitive perception, along with behaviours to achieve more accessible, sociable, hospitable and sustainable cities.

The recent technological advent of solid-state lighting (SSL) and digital control systems is envisioning new possible intelligent applications for urban lighting connected to the wider concept of smart cities. This technological transformation could drive a radical and meaningful change of the contemporary cities at night when adding social value to the urban experience. By exploring the human- and social-scale lighting design criteria, the urban lighting design practice could provide good-quality lighting: visual performances, visual comfort, security and safety perception, visual ambiance and social experiences. To assess these psychological aspects of lighting design from the human and social perspective, this book identifies and compares new tools and research methodologies for knowledge acquisition when tackling with complex issues about the relationship of lighting, people and city.

Book Content

The first section deals with the topic of human and social urban lighting through an extensive literature review that spans from urban, lighting and environmental psychology studies. Chapter 1 reports a brief overview of the importance of public spaces for well-being with the purpose to situate the social dimension of urban lighting in a broader theoretical and cultural context. Chapter 2 presents a critical and systematic literature review about the human- and social-oriented lighting criteria of urban outdoors. Chapter 3 reflects on the methodologies to explore the affective and social appraisals and behaviours of laypersons concerning the lighting design and luminous atmospheres of the nocturnal urban environment.

The second section of the book presents a series of applicative studies elaborated by assessing, analysing, simulating, designing and testing urban lighting parameters, towards an in-depth understanding of the influence of urban lighting in terms of luminous atmosphere perception, positive social affect, social enhancement, accessibility and hospitability of urban lightscapes. Chapter 4 presents a procedure for the analysis, design and post-installation evaluation of an in-situ experience of the human-scale urban lightscape meanwhile Chap. 5 presents the research results related to the influence of lighting on impressions of public outdoor spaces through a virtual experience. Chapter 6 explores two case studies of interactive experience prototyping that has been designed as experimental lighting probes and has been conducted both in the field and in the laboratory. Chapter 7 moves from design research to design practice showing the results of a practice-based/educational experience focused on the development of social-oriented lighting solutions in a specific urban context of application.

Intended Readership

The book aims to extend the knowledge about the spatial experience of human-scale lighting in outdoors, proposing a multidisciplinary qualitative and quantitative methodology to be applied in the lighting design and research and influencing lighting practice and education. In doing so, the book is dedicated to educators that seek to unlock the traditional perspective of urban and lighting design disciplines, to lighting design researchers that need to feed their activities with new methods and inspiration, and to lighting practitioners, both professionals and students, that aim to deepen their knowledge about the human and social dimension of urban lightscapes. The book explores the topic both through a theoretical background, methods and tools and with a research through design approach, to test the theories through applicative research and design examples.

Acknowledgements

Acknowledgements to my past and present colleagues from Politecnico di Milano for their inspiring academic conversations and interesting interdisciplinary research studies together: Eng. Dr. Fulvio Musante for his contagious passion about lighting engineering and Associate Professor Chiara Colombi and Federica Vacca for their trusting encouragement toward the complexion of this book. I would like to thank the Assistant Professors Harm Van Essen and Elke Van Ouden from TU/e

(Intelligent Lighting Institute) for having hosted and collaborated to the lighting pilot conducted at the Markthal Living Lab. My gratitude goes to all the participants of the studies. Finally, I would like to thank warmly my beloved son for his patience during the development of this work.

Milan, Italy Daria Casciani

Contents

Chapter 1
City and Sociality at Night

Abstract This chapter reports a brief overview of the importance of public space for the wellbeing of the urban community. The aim is to situate the social dimension of urban lighting in a wider theoretical and cultural context by describing the spatial anthropology of urban environments. This goal is accomplished through the use of methodologies, theories, and models of public space. Artificial lighting in urban spaces can be an important design lever for fostering the sustainable and inclusive development of territories. This chapter focuses on the paradigm shift that can stem from using artificial lighting as a design tool to recover a sense of place in social urbanity. From this perspective, social planning around urban lighting emerges as an important topic of research and practice. In particular, this new approach to the design of artificial lighting in cities is explored in terms of its aims and benefits, techniques, and methods for fostering inclusive decisions.

Keywords Human-scale city · Social function · Human-scale lighting · Human-oriented lighting · Social inclusion

1.1 The Human and the Social Nature of the City

Since the 1960s, studies in numerous fields, including anthropology, the psychology of perception, and environmental and behavioural psychology, have contributed to our understanding of the connection between architecture, urban space, and the people inhabiting cities. At that time, many theorists began to question the dominance of the car in cities, bringing forward an attitude of human-focused approaches in urban projects. These theorists put forward approaches, methodologies, and metrics for improving the character of urban public space in order to contribute to the well-being of communities.

D. Casciani, *The Human and Social Dimension of Urban Lightscapes*,
PoliMI SpringerBriefs, https://doi.org/10.1007/978-3-030-57165-8_1

1.1.1 The Social Function and Quality of Urban Spaces

The urban design literature stresses the need for meaningful public spaces that can facilitate social interaction. A successful public realm requires spaces where interactions between strangers can occur (Lofland 1999). Pedestrian streets and squares are important for the survival of communities because they could generate, enhance, and sustain a sense of community (Whyte 1984; Gehl 1987; Carr et al. 1995). This has a positive social and psychological impact for the health of modern communities (Jacobs 1961; Oldenburg 1991; Gehl 1987; Lynch 1984). Conversely, an urban space that is devoid of human activity is depressing and uninteresting. This has detrimental economic and social effects. Public spaces have been described as an essential ingredient for the political, social, economic, and public well-being of cities (Banerjee 2001; Florida et al. 2010; Francis 2009). They are useful for the exchange of information, the facilitation of social dialogue, the fostering of social awareness, the enhancement of relationships, the enforcing of local ties, the strengthening of a sense of belonging, and the encouragement of positive ethical conduct. Several authors (Goffman 1967; Hillier and Hanson 1984; Tonnelat 2010) have investigated the social dimension of contemporary cities. Going beyond matters of architecture, they have analysed the city as a social phenomenon made up of relational networks, such as sensations, emotions and gestural interactions, between people and space, and between people and people (Goffman 1967).

Hence, the organization of society is inextricably linked with the organization of urban space: the social use of a space can be examined through the spatial organization of the built environment, and vice-versa, understanding society can help to understand the spatiality of the city (Hillier and Hanson 1984). According to the studies of Hillier, this relation can be assessed through the concepts of integration, connectivity, and accessibility, by quantifying the aspects which spatially organize constructs with social and cultural formation. The psychological, emotional, and behavioural aspects of a place change with the way it is perceived. This human and social layer could shape the activities that take place there. In this regard, Gehl (2010) suggests that urban public spaces should invite people to perform the necessary, the optional, and the social activities within the city. Measures of urban social interaction have been used to determine the level of health, vitality, and liveliness of cities: they are also an indicator of people's satisfaction with their physical surroundings. People prefer to live in public spaces that allow from simple coexistence to conviviality and from passive to active engagement (Banerjee 2001; Mehta 2007; Mehta and Bosson 2018; Florida et al. 2010; Amin 2008; Illich 1974). Indeed, a place may elicit a variety of social relations between people, which could range from passive to active.

Gehl proposes a scale of social intensity that runs from social isolation to close contact with others; its categories include passive contacts, chance contacts, and acquaintances or friends. Successful public spaces provide opportunities for varying degrees of social engagement and disengagement. Passive relationships in an urban situation are based on a "*throwntogetherness*", a situation where people live together in a peacefully situated multiplicity (Massey 2006). In urban settings, where people are unacquainted with each other, the predominant relation is of unfocused interaction through rituals of civil inattention (Goffman 1959). This passive way of relating to others is based on people-watching while avoiding eye contact through an attentive looking away. This "*eyes on the streets capacity*" (Jacobs 1961) is an important measure of the social quality of a space. It indicates the possibility of people watching each other, and it is useful to keep things safe and active. Moreover, it contributes to people feeling safe, observable and so less vulnerable. The best urban spaces provide universal "*watch-ability*" within and along their boundaries. The "*eyes on the street capacity*" can be evaluated with DepthmapX (Software | Space Syntax Network 2019), the open-source software that calculates isovists across a grid throughout the space and indicates areas that provide better views (larger isovists).

Active relationships involve direct experience with the place and with people. People-watching is satisfying because public spaces are not ideal for getting acquainted (Whyte 1980). However, in some cases, people desire more direct interaction with others, both acquaintances and strangers (Gehl and Gemzøe 1996). Thoughtful planning of socially involving public spaces can create opportunities for easy and informal interaction with short-term and low-intensity contacts that results in a positive impact on the social experience. External stimuli can provide linkages between people and prompt strangers to talk to others: engaging experiences may be designed to foster social experiences through unpredictable, varied events in public spaces, which can create a sense of discovery (Lovatt and O'Connor 1995).

1.1.2 Qualities Supporting Successful Urban Social Places by Day

Successful public spaces frequently are meaningful (allowing people to make a strong connection with the space and their personal lives), democratic (protecting the rights of user groups), and responsive (designed to serve the needs of their users) (Carr et al. 1995). Accessibility, comfort, and sociability are the main criteria for a good place. *Accessibility* refers to a place's capacity to bring people together through the affordance of interaction between individual abilities and an environment's spatial organization (Gibson 1986). It can be measured by looking at the physical and visual connections a place has to its surroundings. Affordable and accessible spaces have soft edges and boundaries between inside and outsides, providing shades of transparency to see through: this is inviting and adds interest (Gehl 2010). *Comfort*, the second criterion for a successful public space, can be both psychological (feelings of safety and relaxation) and physical (a pleasant image of the city). Comfortable urban places should engage people in activities, by offering some leisure to the eyes meanwhile avoiding complex or monotonous visual features. Regarding visual aesthetics, Jacobs (1995) talks about magic as a subjective and un-measurable feature of both comfortable and beautiful places. Pleasant urban public spaces are designed in light of the possibilities and limitations of the human body; for instance, environments designed to be experienced at the speed of five km/h for pedestrians provide a rich sensory environment with interesting activities happening at a closer distance (Gehl 2010). The most inviting features of an urban space include scale and rhythm, transparency, appeal to many senses, texture, materials and details, vertical façade rhythm, and mixed functions (Gehl et al. 2006). At sociable places, people meet, invite friends, and feel comfortable when interacting with strangers. According to Jacobs (1995):

> Beyond functional purposes of permitting people to get from one place to another and to gain access to property, streets, most assuredly the best streets, can and should help to do other things: bring people together, help build community, cause people to act and interact, to achieve together what they might not alone.

Indeed, some scholars argue that the social affordances offered by the presence of people in public spaces are more important than the physical affordances offered by the environment (Gibson 1986; Mehta and Bosson 2018).

> Experiencing life in the city is also diverting and stimulating entertainment. The scene changes by the minute. There is much to see: behavior, faces, colors and feelings. And these experiences are related to one of the most important themes in human life: people. (Gehl 2010)

Sociable spaces are defined by the strong attachment of their community, by the social cohesion they provide, by the number of people present and the duration of their stay, by the variety of activities-particularly social activities, and the diversity in age and gender of the users. Sociable places should provide physical opportunities for people to be in the space: a visible, accessible, and public infrastructure has to be provided (e.g. spaces to sit comfortably) (Whyte 1980). The sociability of a space is also connected to safety, which depends on the social field of vision that sets the distance at which it is possible to identify a person by decoding his/her facial expressions.

1.1.3 Human-Scale Urban Public Spaces

Public space has been conventionally defined as an outdoor location that is publicly owned (i.e. streets, parks, and squares), which is in contrast with private houses or the workspace. However, this definition has changed in response to factors such as the growing privatization of public life and the widespread diffusion of the internet. Today, public spaces are "*Hybrid Commons*", defined as semi-public spaces owned and managed by private-public partnerships and accessible to the public. They include internal, external, and external-internal quasi-public spaces (Banerjee 2001). All three types of space are addressed in this book with a particular focus on pedestrians as the main users.

The purest forms of public space include public squares, streets, parks, and parking lots. Streets are the most fitting symbol of the public realm (Jacobs 1995). Human-scale streets are places, itineraries and opportunities for different possible social occurrences: they provide opportunities for short-term, low-intensity contact, and easy interaction with other people in a relaxed, undemanding way (Jacobs 1961; Gehl 1987). Also, sidewalks permit a range of casual interactions between people that "*imply no private commitments*" while protecting one's privacy (Jacobs 1961). Pedestrian-only streets prioritize people walking at five kilometres per hour over vehicular traffic; they are the backbone of the city.

Sidewalks and street-adjacent walkaways serve different functions: they are useful for promenading and performing, and they are also suitable for social stay and leisure activities such as sitting, watching, talking, drinking, eating, dawdling, and playing (Appleyard 1981; Gehl 1987; Carr et al. 1995). Laneways and alleys are narrower streets that give priority to pedestrians, adding diversity to the road network and creating vibrant fine-grained spaces. Pedestrian walkways do not necessarily run along the street; examples might include skywalks (pedestrian overpasses), sub-walks (pedestrian tunnels), walkways giving access to parks, mid-block interiors, and midblock street crossings. Parklets or curbside seating is sometimes set up alongside sidewalks to allow people to gather. Pedestrian squares and plazas are also places of gathering, but they work differently from streets: they allow for social spectacle and cultural or leisure meetings. Public parks embody the presence of nature in the city and offer a place for people to recover from their daily routine. They support sporting activities and recreation; they are also places of refuge from crowds (Jolè 2002). Quasi-public spaces include "*privatised*" areas placed in nominally public and historical environments. Oldenburg calls these "*third places*" to describe

> the great variety of public places that host the regular, voluntary, informal and happily anticipated gatherings of individuals beyond the realm of home and work. (Oldenburg 1991)

Even if legally private, third places like university campuses, shopping promenades, and café terraces form part of the public realm. This book is concerned with the notion of social space in general, regardless of whether any given space is "*public*" or "*public-private*" in nature. Therefore, any outdoor space that supports, enables, or facilitates informal social interaction was considered.

1.1.3.1 Pedestrian Lighting Zones and Activity Levels

Instead of considering the traditional typologies of urban environments, one could consider urban public places in terms of urban lighting functions or people's activity and urban use, as reported into the lighting guides of practice. These guides report that different urban zones such as pedestrian paths, squares, circulation drives, and streets

> should not be illuminated to a common denominator illuminance that conveniently blankets all of them. (IES HANDBOOK, 10th Edition 2011)

Table 1.1 Summary of night-time outdoor lighting zone definitions and activity levels

Urban lighting zones		
LZ0	No ambient lighting	No lighting and adaptation to darkness. Natural environment
LZ1	Low ambient lighting	Low, not uniform ambient lighting for safety and convenience
LZ2	Moderate ambient lighting	Moderate ambient lighting for safety and convenience of users
LZ3	Moderately high ambient lighting	Moderately high, uniform and continuous ambient lighting
LZ4	High ambient lighting	High ambient lighting
Activity levels during night-time hours		
High	Entertainment and shopping districts, transportation hubs, university campuses	High volume of pedestrians during dark hours. Activity level is relative to the local population and activities offered in the outdoor space
Medium	Office and residential complexes, small shopping areas, urban central parks, civic and cultural districts	Moderate volume of pedestrian activity. Small to moderate population size
Low	Residential neighbourhoods, small apartments	Low to very low pedestrian volume with little activity for extended periods. Suburban and rural contexts

On this account, urban areas should be analysed by determining the appropriate night-time lighting zone (from LZ0 to LZ4 as summarized in Table 1.1) in relation to the activity level of people in that area. High activity levels demand greater illuminance; activity level according to time of day demands a suitable level of lighting, in relation to the activity period (IES HANDBOOK, 10th Edition 2011).

The IESNA handbook (2011) and also the CIE Guides (CIE 136 2000) use the following definitions:

- Residential areas: areas of a village, town, or city that are suitable for or are occupied by private dwellings.
- Local roads: complex environments with both pedestrians and cars. Pedestrians are considered the main users of sidewalks or street-adjacent walkways, which are typically paved areas for pedestrians that extend along roadways for vehicular traffic.
- Commercial areas and shopping promenades: settings designed specifically to attract shoppers to the area; they are generally located in the city centre where buildings and monuments of historical or local importance can be used to stimulate people's curiosity.

- Pedestrian walkways that are street distant and do not necessarily run along the street; examples include skywalks (pedestrian overpasses), sub-walks (pedestrian tunnels), and walkways giving access to parks, midblock interiors, or midblock street crossings.

1.2 Urban Planning After Dark: Night-Time Lighting Design

The theories mentioned above have one important limitation: few consider the night-time conditions of urban public spaces. Lighting has been an afterthought in these accounts; it is usually treated as part of the functional infrastructure of the city or a standard element of street furniture (Carmona et al. 2002). Few accounts have recognised the importance of lighting as a medium that adds quality to public spaces and could affect night-time sociability. Jacobs, for instance, considers how lighting can support the social life of urban spaces and create visual qualities that engage the eyes.

> What of streets at night? They may almost cease to exist [...] or they may exist only at night, because of the light and what it does to the eyes. (Jacobs 1995)

He specifically focuses on streetlights as common urban elements that can be considered a special design feature that contribute to the quality of streets in terms of safety and accessibility (Jacobs 1995). Gehl stresses the importance of supporting the social field of vision at night through carefully designed lighting:

> Lighting is crucial once night falls. Good lighting on people and faces and reasonable lighting for facades, niches and corners is needed along the most important pedestrian routes to strengthen the real and the experienced sense of security, and sufficient light is needed on pavements, surfaces and steps so that pedestrians can maneuver safely. (Gehl 2010)

The importance of lighting goes beyond being a functional element that ensures safety, comfort, and accessibility along with city beautification. This book focuses on human scale lighting as a key factor for fostering the social dimension of public spaces after dark.

1.2.1 The Importance of Urban Lighting

The increasing number of people leaving the countryside and moving to urban areas has led to a situation in which more than 68% of the world's population will live in cities by 2050 (United Nations, Department of Economic and Social Affairs, Population Division 2018). This trend poses questions about how to ensure suitable living conditions in the cities and how to guarantee the energetic provision for all. Urbanisation is one of the main drivers of change for urban lighting. Urban lighting is an important factor not only in terms of energy consumption and cost for municipalities, but it is also crucial for creating the conditions that allow for the positive use of the city after nightfall (Boyce 2019). The traditional idea of urban public lighting is based on two antithetical positions: on one side, economic savings and technical performance based on quantitative measures; this position is well depicted by lighting regulations aimed at energy savings and safety provision (UNI EN 13201-2 2016; UNI 11248 2016). On the other side, lighting contributes to urban amenity city beautification and that can stimulate night-time activity. This second approach guarantees the pleasantness and attractiveness of cities after dark: lighting encourages the use of the commons during the night (Fotios et al. 2017), enables citizens to have a positive social life, and contributes to the economic health of twenty-four-hour cities. In addition, lighting has become a touristic attraction through the organization of light festivals that transform the entire city into a temporary open-air museum (Giordano and Ong 2017).

1.2.2 The Human and Social Scale Lighting in the Urban Night-Time

Historically, artificial light has changed considerably the way that people experienced the city at night (Brandi et al. 2007). A critical, simple, and obvious question may arise for the reader: given its definition, doesn't all urban lighting design have a social dimension? The answer is not as banal as might be expected. Outdoor lighting is primarily intended for people. Therefore, urban lighting should be focused on the human experience, particularly in terms of facilitating social engagement (Casciani and Rossi 2012). As previously described, lighting design in the urban context has two main objectives: functional lighting intended to meet safety needs through code compliance, and architectural, effect, or event lighting intended to highlight the architecture of the city and achieve an aesthetic effect. Few studies and recommendation from the CIE (Commission International de l'Eclairage standing for International Commission on Illumination) agree that urban lighting is primarily intended for people's experiences, activities and needs in the urban space (CIE 136 2000; CIE 115 2010) This position is also supported by lighting designer and practitioners (Brandi et al. 2007; Raynham 2007; Raynham and Gardner 2001). They state that urban lighting should ensure equal and free access to the city after dark, both in the centre and in suburban areas, for all citizens, regardless of their social status.

> Lighting should contribute to reducing social and economic inequalities and support integration policies. (Charter on Urban Lighting—LUCI Association 2019)

In the last ten years, there has been emerging interest in human-oriented lighting performance and social lighting in the city, above all, through lighting design practice. Besides official sources, lighting movements have arisen, blending urban lighting practices with social discipline. In 2009, the lighting designer Roger Narboni declared that architectural lighting was dead and that lighting designers should do more to address people's needs rather than just highlighting architectural monuments.

> To an era of strictly functional public lighting followed by a period where using lighting as tools of heritage beautification developed rapidly. Wouldn't it now be time to think of the users and to invent what we could call a social light, meaning a light and ambiance responsive to the needs, the experiences and emotions of the users and citizens? (Narboni 2012)

From this perspective, the human and social dimension of urban lighting is important for providing comfortable and accessible environments, for providing a sense of safety, for contributing to the ambiance of the city, and for enhancing the sociality of the city after the sun has gone down. Urban lighting can facilitate social cohesion, encouraging citizens to congregate in an inclusive way that encourages civic participation.

More than this, these approaches can ensure that lighting is adapted to the activities and the needs of local people for ensuring safety but also lighting events and celebrations. It was in this spirit that the idea of the *shades of night* (Arup 2015) was conceptualized: the shades correspond to the uses and needs of different public spaces at different times. The idea is that lighting should respond to the beat of the city in its micro-areas and local communities, reflecting the true meaning and purpose of public spaces. Responsive lighting is intended to support the uses, activities, and needs of people in order to develop a more reflective and ecologically responsible way of using artificial light (Tillett 2011). Lighting can make a small but important difference in people's lives. Lighting rarely leads people to have a big emotional rush, but it positively, sometimes unconsciously, affects people's lives.

The philanthropic Social Light Movement suggests another aspect of the human and social lighting approach. The group aims to improve lighting for people who are unlikely to have access to good quality illumination. In their manifesto, they claim that "*lighting is a right, not a privilege*" and involve communities in the lighting design of their own environment. The goal is to provide good quality lighting to create attention and gain the support of public opinion while promoting responsible energy use. They encourage the integration of people into the lighting design process to better understand how lighting can be responsive to people's needs (Social Light Movement 2019).

From these different approaches, it is evident that the life of cities after dark needs to be addressed from a wider social, economic, and environmental perspective (Fig. 1.1). A human and social-oriented approach to lighting is emerging in research and practice as a young and promising field. This field is focused on providing citizens with equal access to good quality lighting in order to contribute to the sustainable and inclusive of the public realm at night (Casciani and Rossi 2012). Indeed, the relationship between lighting and people's experience reflects new ecologies of urban lighting. Urban lighting can enable inclusive and equitable use of public spaces: it could act as a cost-effective and sustainable socio-technical

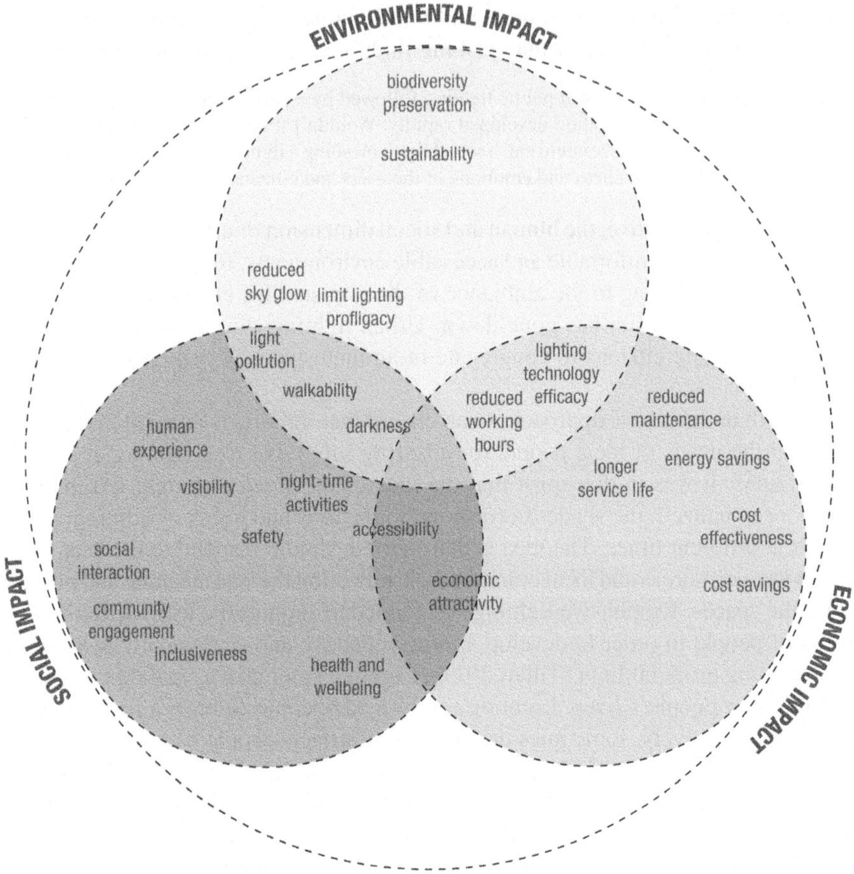

Fig. 1.1 Three-part approach to human and social scale lighting considering the social, environmental, and economic impact on the urban night

infrastructure that can create engaging, accessible, and comfortable urban areas for the diverse citizens that share them (Sloane et al. 2016).

In comparison to energy, technical, and economic issues, the social and experiential value of urban public lighting is difficult to estimate because the existing literature is focused on indoor lighting applications. In addition, research on the effects of lighting on social interaction is rarely considered when policy decisions are made. As a result, we lack comprehensive studies on the topic and have a limited sense of a shared research methodology. Lighting design research on the perceptive features of lighting (sensorial, affective, cognitive, and evaluative) requires a user-centred and multidisciplinary approach: theories and models from the social sciences, behavioural sciences, and environmental psychology are fundamental to understand the importance of lighting for people. This approach ensures that the focus on reducing energy use is not detrimental to the human experience of cities after dark; similarly, it ensures that lighting technologies are designed to enhance citizens' conviviality, sociability, and relational trust (Tillett 2011; Bordonaro and Aghemo 2006; Bessette 2011).

1.2.3 The Importance of Urban Darkness

Concerns about light pollution have grown. Urban lighting can affect ecological systems, wasting energy and disrupting the natural behaviour of animals and humans. Light pollution at night is problematic due to its impact on human health and the ecosystem: it disrupts both the biodiversity of urban environments and natural ecological processes such as bird migration, animal reproduction, predation and foraging, and human circadian rhythms (Navara and Nelson 2007). The overuse of artificial light at night reduces the visibility of the stars by limiting the night's visual contrast, partially as a result of sky glow (Lighting Research Center 2003). Sky glow is a combination of reflected and refracted light, determined by the cumulative effects of street lights, outdoor advertising, lit buildings, parking lots, airports, and other lighting sources. The phenomenon of light pollution can also take other forms: for instance, glare is the creation of excessive contrast between bright and dark areas in one's field of view. It results in difficulty seeing because of the impact of direct or reflected bright light in the field of view. Finally, light trespass is unwanted light that spills onto private property (e.g. light that extends onto the facade of buildings, into windows, or into the sky).

These phenomena are contributing to the absence of natural night in cities. Although there are benefits of illuminating outdoor spaces at night, astronomers and biologists claim that the effects of the over-use of bright lighting are devastating. Hence, around 2010, a new school of thought in urban planning emerged that centred on the idea that "*less lighting is more*", subverting the previous approach of "*the more the better*". The "*dark infrastructure*" methodology in urban planning has been defined to re-establishing natural night-time conditions to restore nocturnal biodiversity (Zielinska-Dabkowska 2019). This lighting design approach focused on determining the right lighting level based on the connection between people, light, and nature (The Royal Commission on Environmental Pollution 2009).

Darkness is often described as a natural resource at risk in terms of design possibilities and urban nocturnal expressivity:

> I think the problem we now have is that it is too bright outdoors. You need shadows to express light […]. Some cities are so bright that there are no dark corners anymore. (Laganier and Pool 2011)

An attitude of environmental sustainability, along with an emphasis on the economy, supports the trend of keeping cities dark at night; the goal seems to be to benefit living species and reduce consumption. The dark age of the cities appears to have been welcomed by public institutions that need to cut public costs. A rational approach for decreasing lighting levels would entail providing lighting when and where it is needed. For instance, the efficacy of street lighting is determined by the selection of the most suitable lighting source for that situation in combination with the appropriate lighting fixtures accompanied by proper management and maintenance. The approach of upgrading lighting systems with more efficient technologies that provide better lighting distribution could reduce glare, light trespass, and sky glow. Conversely, issues of sustainability cannot be addressed only by replacing one technology with a more efficient one (Owen 2010). The reduction of lighting could be supported through the use of reflective, refractive, and photo-luminescent materials; these allow lighting to be manipulated at a human scale to better define visual boundaries and manage visual cues (Tillett 2011; Zardini 2005), creating a setting that is interesting at night. In addition, the purposeful timing of lighting that considers the real activities of people could dim nightscapes without having detrimental effects on the experience of the city after nightfall (Seitinger and Warwick 2013; Kohler and Sieber 2011). A rational approach to energy consumption in outdoor lighting requires the right light, for the right place, at the right moment. In this regard, digitized and programmable LED lighting can be used to address local and social needs. In this view, lighting in the city is a dynamic, lively, and active element that evolves throughout the night. If a variable lighting system is suggested for security and energy-saving purposes in response to the needs of traffic usage variation (BS 5489-1 2013), it can be also used for signalling and accentuating certain events or for supporting social activities. Unfortunately, the implementation of this approach would encounter several obstacles: the reluctance of elected officials, the presence of conflicting regulations, and above all, users' discomfort with the approach. In this direction, it is envisioned the necessity to include laypersons in the assessment of new lighting schemes at night.

1.2.4 The Role of the Designer of the Urban Night

The Urban Lighting Master Plan is a holistic tool and comprehensive planning document introduced about 40 years ago; it proposes an approach to the night-time image of several French cities (Zielinska-Dabkowska 2019). Even if the Urban Lighting Masterplan has not served as the basis for international regulations or rules of practice, it can be used as an instrument to guide the design of the urban night. It aims to define a systematic and coordinated framework for artificial lighting that addresses the environmental, historical, cultural, and social elements of the urban context. The traditional approach of urban planning is a top-down decisional process in which municipalities select an urban lighting designer, architectural lighting designer, and urban lighting planner to choose and implement a lighting strategy. Conversely, the more complex of a solution that is called for, the wider and more complete a project team needs to be, involving different skills, methodologies, and know-how (Zielinska-Dabkowska 2019). A wide number of professionals and disciplines are required to define innovative urban lighting projects that will be accepted by the community. This will allow the team to design technologically up-to-date solutions based on people's lighting experience (Gentili and Casciani 2016a, b).

If the crucial aim is to fulfil the needs of citizens which are considered the most important clients of the lighting masterplan, lighting designers should collaborate with social scientists, anthropologists, and ethnographers so that they can be sure that they are addressing citizens' needs. A multidisciplinary approach can contribute to achieving inclusive community lighting decisions. This bottom-up approach might entail the collection of data through door-to-door canvassing, interaction with local spokespersons and networks, or contact with local associations. The idea is to build a solid connection with residents at the micro-scale of the neighbourhood. The attitude of the lighting designer towards the people for whom the lighting design is intended is crucial for the smooth functioning of the project. Listening to inhabitants allows the designer to identify the needs of the local population. This link is how a lighting design solution should be communicated to, and evaluated by the city's inhabitants.

Some cities have already proposed social lighting as a strategy of their urban masterplan. Since 2007, the city of Lyon has used the Lighting Masterplan to inform attempts to survey and incorporate residents' feedback on an overall night-time lighting strategy. This strategy integrated new technical possibilities that increased the efficiency of the public lighting system while also taking into account how residents made use of urban public spaces. The inclusive lighting development was focused on adapting to activities and rhythms of the citizens. More recently, several cities, such as Amsterdam, London, Paris, Zurich, New York, and Tokyo have appointed Night Mayors. These are influencer figures focused on managing and improving relations between city government, city residents, and night businesses (Night Mayor 2019). Healthy cooperation and dialogue between the City Council and urban nightlife participants can minimize conflict and create a stimulating night-time atmosphere.

1.2.4.1 Tools, Techniques and Methods for Community Involvement in Urban Lighting

Lighting design practice commonly requires study of the territory and urban context (developed in the Lighting Masterplan) based on:

- a historical analysis of the transformation of the structure and dimensions of the space, the evolution of the identity of the place, and changes in terms of urban lighting that occurred in that specific zone;
- a state-of-the-art analysis of the morphology of the space and its urban lighting system (fixtures and plant).

Analysis of the relationship between space/context/urban structure and people is less common, though preferable. If this analysis is normally generalized in a way that takes for granted that certain spaces are used for certain functions, then a deeper investigation into the specificity of certain places is needed for producing better design solutions. Urban lighting is now conceived as a public service with lighting design seen as an ethical issue. The urban lighting designer should play the role of a night-time experience designer who is competent in the three areas of technology, human factors, and design. The designer must integrate multiple disciplines to explore how urban lighting impacts human experience and transform the urban space (and vice-versa). Since the human and social value of lighting is hard to estimate, designers should draw upon theories, methods, and models from social science, behavioural science and environmental psychology to assess the intangible aspects of the link between people and lighting (Bordonaro and Aghemo 2006; Tillett 2006, 2011; Gentili and Casciani 2016a, b; Bordonaro et al. 2019). New methods, tools, and techniques should be explored and tested for community involvement in urban lighting (Fig. 1.2).

The designer could start from observation, becoming a detective to interpret the night-time landscape in terms of human and social experience. This might entail that people be observed and monitored in their real-life context. The Nocturnal Flaneur

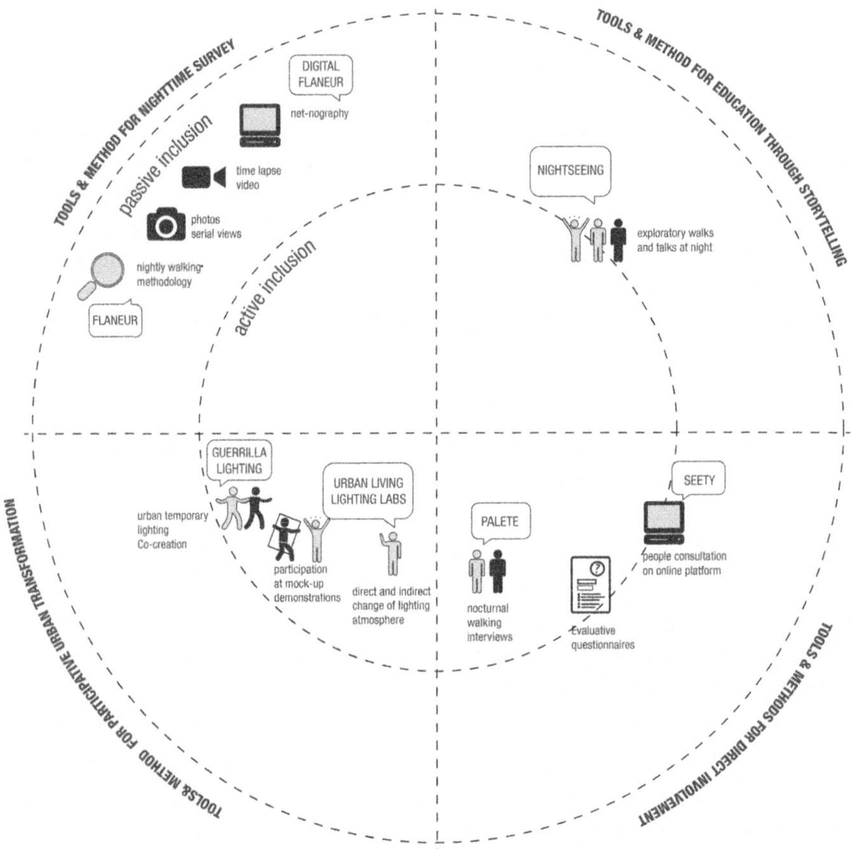

Fig. 1.2 Methods, tools and techniques for passive and active community inclusion in urban lighting design

approach works by taking the point of view of pedestrians through a nightly walking methodology that allows the researcher to see the space through the eyes of people and to document it through photography and video. These documents represent the nocturnal visual experience. The Flaneur could use video-monitoring tools to make observations. This can be achieved remotely with video cameras able to acquire real-time data about the use of the city after dark (Köhler and Sieber 2011). The digital Flaneur can also search online forums to study virtual discussions generated by the community. This practice is based on the virtual ethnography or net-nography approach (Kozinets 2010). These techniques will reveal residents' habits during the hours of observation. A more comprehensive lighting strategy can be developed based on these data.

Other approaches entail the deeper involvement of citizens in the development of the project. CERTU (Organisme d'études du Ministère français de l'Urbanisme) and LUCI (Lighting Community urban association) argue for the necessity of focused communication with the public and the need to develop a dialogue with community associations that will include citizens in the design phase. In this, lighting designers are committed by social and ethical obligations to be inclusive. The designer could become a walking educator and an urban lighting storyteller by exploring the nocturnal territories and involving people in exploratory walks at night. These could be aimed at discovering the urban night-time setting while engaging communities in critical reflection on the issue of lighting. The NightSeeing™ (NightSeeing 2019) is an example of a situation in which local inhabitants and tourists are incentivized to discover the city after dark and to critically reflect on their surrounding environment. The advantages of storytelling rely on the fact that the impressions and experiences of participants can be collected and informally exchanged. The proposed night walks raise awareness of the profession of lighting design and demonstrate the importance of artificial light. Consultation with people is an uncommon approach for social inclusion in lighting design projects; however, information acquired through this method is grounded in a specific community and can inform both lighting designers and decision makers. In that case, the designer directly involves people as advisors through public outreach, employing evaluation techniques such as questionnaires and interviews. The goal is to gather information about people's perception of the lit environment and also to motivate people to reflect on their surroundings.

Another technique of inquiry is PALETE (Parcours Libre et Tèlè-Enregistès), which was implemented at the Centre de Documentation du laboratoire Cresson (Fiori et al. 2004). These nocturnal interviews aim to understand the sociological aspects of urban lighting by collecting data on the perceptions of residents, which can then inform strategies to enhance lighting design projects. During these walks, residents and authorities are interviewed and invited to describe and comment on the problems of the neighbourhood, as well as their perception of urban lighting in general (both in functional and aesthetic terms). More than this, they are asked to give suggestions for improvement. Evaluative questionnaires about the perception of urban lighting can be distributed to define a shared and polyphonic lighting project (Deleuil 2009). Wider involvement is possible through the use of online public platforms that allow citizens to give their opinion about the lighting in their neighbourhood or their city: an example includes SEETY Milano (SEETY 2019; Casciani 2014), which enables participants to vote on, select, or refuse a project; they can also propose ideas, solutions, and possible alternatives. The general requirement for these surveys is that the participants be active and that the platform be reachable, open, understandable, and used correctly. Even though these platforms of public outreach are rarely used in critical decisional projects, they are a useful tool for gathering insights into and impressions of nocturnal territories and for engaging people in discussion about lighting.

The designer could help transform and renew the nocturnal territory. This could happen through an approach of urban lighting co-creation with people who are impacted by the renewal of their nocturnal territory. Such a process might entail the preparation of lighting performances, lighting fixtures, and lighting settings. With the support of professionals, people would express their belonging, imagination, and memory in relation to space, which is generally a neglected and abandoned territory. This would then be renewed through lighting. People can also attend meetings where prototypes, mock-ups, and demonstrations are presented. A communication strategy should be implemented to update residents about the lighting transformation of their neighbourhoods. Although participation during the installation phase is burdensome in terms of time and resources, pilot tests are useful to give residents an idea of the lighting projects proposed, which would provide experts with an opportunity to gather comments and useful information, as well as to modify or improve the project (Brandi et al. 2007). Lighting workshops are another effective method to understand people's experiences of the urban night and, at the same time, help people understand the urban environment. These workshops are generally organized as temporary events in the form of guerrilla lighting (Social Light Movement 2019) where students and residents work together to change temporarily the lighting atmosphere of a space through the use of portable lighting systems. The transformation of the nocturnal environment, today, can be achieved with digital technologies that can be directly manipulated by users who can choose from a range of lighting atmospheres. In this scenario, people take an active part in shaping the urban environment by changing lighting features and working with other people who enjoy the activity. Urban Living Labs can be a catalyst for innovation and co-creation in cities by testing innovations in a process that involves citizens, businesses, and public authorities. The Valotarina (Light Stories) Project is an example of this, which enables city dwellers to impact the atmosphere of the streets at night through digital storytelling by posting online recordings or sharing stories, messages, or greetings (Pihlajaniemi et al. 2012). In this project, lighting sets the scene to create an experience or shape a space, which people then help to define by taking partial control of the lighting system. Experiential lighting is an important tool, but it is under-investigated. Public feedback observations and evaluations are important to create a welcoming mood, to define a social experience, and to create an attractive night-time urban scenario.

References

Amin A (2008) Collective culture and urban public space. City 12(1):5–24. https://doi.org/10.1080/13604810801933495

Appleyard D (1981) Livable streets. University of California Press, Berkeley and Los Angeles, CA

Arup (2015) Cities alive. Rethinking the shades of night. London

Banerjee T (2001) The future of public space: beyond invented streets and reinvented places. J Am Plann Assoc 67(1):9–24. https://doi.org/10.1080/01944360108976352

Bessette S (2011) Urban lighting 2.0: towards a convivial, animated city. In: 3rd PLDC. VIA Verlag, Madrid

Bordonaro E, Aghemo C (2006) Urban lighting: a multidisciplinare approach. In: International conference proceeding the urban nightscape, Athens, Greece

Bordonaro E, Entwistle J, Slater D (2019) The social study of lighting. In: Davoodian N (ed) Urban lighting for people: evidence-based lighting design for the built environment, 1st edn. RIBA Publishing, London, pp 3–17

Boyce P (2019) The benefits of light at night. Build Environ 151:356–367. https://doi.org/10.1016/j.buildenv.2019.01.020

Brandi U, Geissmar-Brandi C (2007) Light for cities. Lighting design for urban spaces: a handbook. Springer, Berlin

BS 5489-1 (2013) Code of practice for the design of road lighting. Lighting of roads and public amenity areas

Carmona M, Heath T, Oc T, Tiesdell S (2002) Public places—urban spaces: the dimension of urban design, 1st edn. Elsevier/Architectural Press, Amsterdam

Carr S, Francis M, Rivlin L, Stone A (1995) Public space. Cambridge University Press, New York

Casciani D (2014) SEETY Milano: strumenti di partecipazione all'illuminazione urbana. In: Vitellio I (ed) Città open source. Spazio Pubblico - Network - Innovazione Sociale. INU, Istituto Nazionale di Urbanistica, pp 212–214

Casciani D, Rossi M (2012) ELSE, Experience of lighting sustainability in the environment. In: Cumulus, "Open, participative cities: how design knowledge can support public services in the development of open, participative human environment". Helsinki

Charter on Urban Lighting—LUCI Association (2019) Retrieved from https://www.luciassociation.org/about-luci/charter-on-urban-lighting/

CIE 115 (2010) Lighting of roads for motor and pedestrian traffic

CIE 136 (2000) Guide to the lighting of urban areas

Deleuil J (2009) Eclairer la ville autrement. Presses polytechniques et universitaires romandes, Lausanne

Fiori S, Houdemont K, Leroux M (2004) Parcours nocturnes: étude sociologique sur l'ambiance et les usagers nocturnes Mail d'Empalot et cité Bordelongue (Toulouse). Rapport de recherche 61 hal-01364547f, Cresson, p 64

Florida R, Mellander C, Stolarick K (2010) Beautiful places: the role of perceived aesthetic beauty in community satisfaction. Reg Stud 45(1):33–48. https://doi.org/10.1080/00343404.2010.486784

Fotios S, Uttley J, Fox S (2017) A whole-year approach showing that ambient light level influences walking and cycling. Lighting Res Technol 51(1):55–64. https://doi.org/10.1177/1477153517738306

Francis M (2009) Convivial urban spaces: creating effective public places Henry Shaftoe. J Urban Design 14(3):404–405. https://doi.org/10.1080/13574800903057083

Gehl J (1987) Life between buildings: using public space. Reinhold, New York

Gehl J (2010) Cities for people, 1st edn. Island Press, Washington, D.C.

Gehl J, Gemzøe L (1996) Public spaces—public life. The Danish Architectural Press, Copenhagen

Gehl J, Kaefer L, Reigstad S (2006) Close encounters with buildings. URBAN DESIGN Int 11(1):29–47. https://doi.org/10.1057/palgrave.udi.9000162

Gentili H, Casciani D (2016a) Envision the nocturnal territory. Pad Pages Arts Design 13(Design for territories):99–119

Gentili H, Casciani D (2016b) Conceiving the nocturnal city through strategical and tactical urban lighting practices. In: Proceedings of the international conference "A new cycle of urban planning between tactics and strategy". Planum Publisher, Milano, pp 310–317
Gibson J (1986) The ecological approach to visual perception. LEA, London
Giordano E, Ong C (2017) Light festivals, policy mobilities and urban tourism. Tourism Geographies 19(5):699–716. https://doi.org/10.1080/14616688.2017.1300936
Goffman E (1959) The presentation of self in everyday life. Anchor Books, New York
Goffman E (1967) Interaction ritual: essays on face-to-face behavior, 1st edn. Pantheon Books, New York
Hillier B, Hanson J (1984) The social logic of space. Cambridge University Press, Cambridge
IES handbook, 10th edn (2011)
Illich I (1974) La convivialità. Mondadori
Jacobs J (1961) The death and life of Great American cities. Random House, New York
Jacobs A (1995) Great streets. The MIT Press, Cambridge
Jolè M (2002) Espaces publics et cultures urbaines: actes du séminaire du CIFP de Paris, 2000-2001-2002. CERTU, Lyon
Köhler D, Sieber R (2011) The time factor in relation to the electric lighting in urban spaces. Prof Lighting Design 79:38–42
Kozinets R (2010) Netnography: the marketer's secret weapon. White paper. Retrieved from https://www.technologyreview.com/s/421208/netnography-the-marketers-secret-ingredient/
Laganier V, van der Pool J (2011) Light and emotion, exploring lighting cultures conversation with lighting designer. Birkhauser Architecture
Lighting Research Center (2003) Implementation of decision-making tools that address light pollution for localities planning street lighting. Whyte paper efficient street lighting and light pollution
Lofland LH (1999) The public realm: exploring the city's quintessential social territory, 1st edn. Aldine de Gruyter
Lovatt A, O'Connor J (1995) Cities and the night-time economy. Plan Pract Res 10(2):127–134. https://doi.org/10.1080/02697459550036676
Lynch K (1984) Good city form. MIT Press, Cambridge
Massey D (2006) For space. SAGE, London
Mehta V (2007) Lively streets. J Plan Educ Res 27(2):165–187. https://doi.org/10.1177/0739456x07307947
Mehta V, Bosson J (2018) Revisiting lively streets: social interactions in public space. J Plan Educ Res. 0739456X1878145. https://doi.org/10.1177/0739456x18781453
Narboni R (2012) Ambiance et atmosphère en lumière urbaine. Lecture, Nantes
Navara K, Nelson R (2007) The dark side of light at night: physiological, epidemiological, and ecological consequences. J Pineal Res 43(3):215–224. https://doi.org/10.1111/j.1600-079x.2007.00473.x
Night Mayor (2019) Retrieved from https://www.citylab.com/solutions/2016/01/night-mayor-amsterdam-mirik-milan/433893/
NightSeeing (2019) Retrieved from https://www.nightseeing.net/
Oldenburg R (1991) The great good place. Marlowe & Company, New York
Owen D (2010) The efficiency dilemma. Annals of environmentalism, The New Yorker. Retrieved from https://www.davidowen.net/files/the-efficiency-dilemma.pdf
Pihlajaniemi H, Österlund T, Luusua A, Tanska T (2012) LightStories: designing and evaluating interactive and participatory urban lighting. In: Interactive design light workshop, Glasgow
Raynham P (2007) Public lighting in cities. In: International conference Illuminati, Cluj-Napoca, Romania
Raynham P, Gardner C (2001) Urban lights: sustainable urban lighting for town centre regeneration. In: Lux Europa 2001, Reykjavik
Social Light Movement (2019) Retrieved from https://sociallightmovement.com/
Software | Space Syntax Network (2019) Retrieved from https://www.spacesyntax.net/software

SEETY (2019) Retrieved from https://seety.altervista.org/
Seitinger S, Warwick J (2013) Timing is everything: from dynamic lighting to meaningful experience. CHI, Paris, France
Sloane M, Slater D, Entwistle J (2016) Tackling social inequalities in public lighting. Configuring Light/Staging the Social Research Programme, London
The Royal Commission on Environmental Pollution (2009) Artificial light in the environment. TSO (The Stationery Office), UK
Tillett L (2006) A social approach to urban lighting. In: International conference proceeding the urban nightscape, Athens, Greece
Tillett L (2011) Grounding practice: speculations on affect and environment. In: 3rd PLDC. VIA Verlag, Madrid
Tonnelat S (2010) The sociology of urban public spaces. In: Hongyang W, Michel S, Guofang Z (eds) Territorial evolution and planning solution: experiences from China and France. Atlantis Press, Paris, pp 84–92
UNI 11248 (2016) Illuminazione stradale - Selezione delle categorie illuminotecniche
UNI EN 13201-2 (2016) Road lighting. Part 2: Performance requirements
United Nations, Department of Economic and Social Affairs, Population Division (2018) World urbanization prospects: the 2018 revision, key facts. Working paper no. ESA/P/WP.252. United Nations, New York
Whyte W (1980) The social life of small urban spaces. Project for Public Spaces Inc., New York
Whyte W (1984) Learning from the field. A guide from experience. Sage Publication, California
Zardini M (2005) Sense of the city. Canadian Centre for Architecture—Centre Canadien d'Architecture, Montréal
Zielinska-Dabkowska K (2019) Urban lighting masterplan—origins, definitions, methodologies and collaborations. In: Davoodian N (ed) Urban lighting for people: evidence-based lighting design for the built environment, 1st edn. RIBA Publishing, London, pp 18–41

Chapter 2
The Human-Scale Urban Lighting Experience

Abstract This chapter presents a critical and systematic literature review about the topic of urban lighting design explored from the perspective of people. Starting from the mechanism of nocturnal physiological human vision to perception and cognition, the chapter is aimed at systematizing the knowledge about human factors in urban lighting by highlighting the relation between perceived attributes of lighting and the experience of outdoor spaces. The human-scale urban lighting experience highlights the understanding of the interaction between the combined physical external stimuli (i.e. artificial lighting, urban volumes, surfaces and material) and the physiological, cognitive, affective, interpretative and evaluative mechanism of perception of human beings. The human and social-oriented criteria of lighting for the perception of urban outdoors are presented and described to evidence five macro-categories that explore the lighting quality for pedestrians: functionality (accessibility, spatial affordance, visibility, visual comfort, visual acuity, glare, facial recognition, urban legibility, spatial recognition, visual orientation, way-finding, saliency and visual hierarchy of perception), wellbeing (security, safety perception, reassurance), ambiance and urban beautification (architectural, material and landscape enhancement, socially inclusive/inviting, relaxing, interesting and expressive, arousing and lively luminous atmospheres), relationship/experience (personalization, socialization, interaction), and environmental issues (skyglow reduction, light profligacy avoidance and light trespass avoidance).

Keywords Experience · Sensation · Perception · Human-scale lighting · Functionality · Wellbeing · Environment · Ambiance · Socialization

2.1 The Visual Experience of the Urban Night-Time

The lighting design practice has the important role of creating lighting schemes which positively influence the perceived environment and impact human evaluations, emotions, and behaviours, especially in the urban night-time (Cuttle 2008).

D. Casciani, *The Human and Social Dimension of Urban Lightscapes*, PoliMI SpringerBriefs, https://doi.org/10.1007/978-3-030-57165-8_2

2.1.1 *From Sensation to Perception of the Luminous Environment*

Light and illumination are inseparably linked to the sense of sight. All human activities are around 80% dependent on information gathered as visual stimuli by the eyes and elaborated by the complex and rich visual apparatus that lay behind. The luminous environment generates the retinal image that is the stimulus for the process of vision to start, providing information to enable the visual perception process (Cuttle 2008). The visual system is based on the eye and brain working together for processing (sensing and making sense) the images gathered from the environment through the combination of sensation and perception.

Briefly described, the first phases of the visual system are physical: light passes through the transparent protective layer of the cornea and through the lens. Depending on the amount of lighting, the iris contracts or expands (retinal accommodation). The derived visual image is projected upside down in the back of the eyeball where four different kind of photoreceptors compose the retina. Conventionally grouped in two main classes, rods and cones translate the stimulus of light in nervous impulses: the retinal image is transduced into electrochemical neuronal activities that code the characteristics of the sensory stimulation. Signals from the retina are then transmitted to the visual cortex of the brain over the central visual pathways: the optic nerves reach the optic chiasm and the lateral geniculate nuclei to supply information to various parts of the visual cortex, where vision occurs (Schreuder 2008). In this path, the

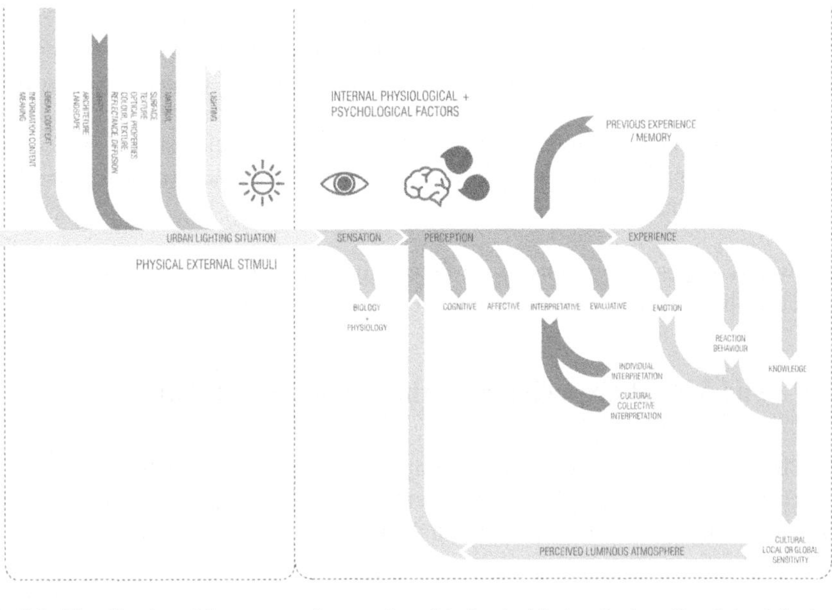

Fig. 2.1 Visualization of the process of perception with physical factors (external) and physiological and psychological factors (internal)

visual stimuli is processed into sensation. Perception occurs through the collection and organization of visual sensations through higher order brain activities that shape the meaningful interpretations of the perceived environment. The physical message is interpreted and codified through the interpretation of certain regularities and variables of the luminous phenomena: the intensity or the perception of lightness, the spectral distribution emitted or reflected by objects and surrounding environment, the lighting distribution (lightness and darkness composition) and perceptual constancies (lightness, colours, shape and size) of perceived objects and environments (Fig. 2.1).

2.1.1.1 Vision at Night

Cones and rods enable the visual system to work in different situations, by daytime and night-time. Rod photoreceptors contain the same photo-pigments with the same spectral sensitivity. Cones have three different photo-pigments and a relative spectral sensitivity functions: short (S), medium (M) and long (L) wavelength cones. Rods and cones are also different in quantity, functions and disposition. Rods are uniformly distributed across the retina and contribute to peripheral vision, allowing to see in a 70° conic field of view. Being a very old visual system in evolutionary terms, they are very sensitive to low illumination levels, to blue-green light, to contrast and motion detection, helping human beings to perceive the nocturnal environment (Schreuder 2008; The SLL Lighting Handbook 2009). Cone photoreceptors are concentrated in the fovea and are mainly responsible for colour discrimination and for detailed vision in about a two-degree cone of the visual field. Human vision adapts in a broad range of light and dark conditions from near complete darkness (10–6 cd/m^2), to full bright sunlight (106 cd/m^2), operating in:

- *Photopic vision* is based on cones and occurs in bright light conditions, (luminance levels 1–106 cd/m^2).
- *Scotopic vision* is based on rods and occurs at very low light (less than 0.001 cd/m^2)
- *Mesopic vision* occurs in semi-dark conditions (between about 0.001 and 3 cd/m^2), being a combination of photopic and scotopic vision.

In urban environments the visual system operates in mesopic vision (The SLL Lighting Handbook 2009). During night-time conditions, the processes of visual adaptation from darkness to brightness (light adaptations) and vice versa (dark adaptation) are really important for human vision and perception of the urban space, considering the variety of urban lighting composition from city centers to suburban areas and also from bright to dark areas in the same urban zone. The timing for dark adaptation is about 20–30 min, while, the light adaptation takes around 5 min. In case of urban environments and mesopic condition, dark adaptation requires a shorter time.

2.1.2 The Psycho-social Perception of Cities

Based on a complex perceptual process of understanding of the sensorial stimuli gathered from the environment, the psycho-social perception of "*urbanscapes*" operates simultaneously on four dimensions to organize and make sense of information (Bell et al. 2001; Rapoport 1977):

- *Affective perception* involves the emotional side of environmental perception;
- *Cognitive perception* involves thinking about, organizing and keeping information, making sense of the environment and creating mental map;
- *Interpretative perception* encompasses meaning or associations derived from the environment, relying on memory (prior experiences) to compare with new experience, thus activating expectations;
- *Evaluative perception* incorporates values and preferences such as the determination of "*good and bad*".

All these aspects of perception are inextricably interwoven in the real and total experiences of the environment. Light and lighting design send visual messages through the perceptual pathway, influencing human affective, cognitive, interpretative and evaluative responses (Steidle and Werth 2014; de Kort and Veitch 2014; Light + Behavior Symposium—Illuminating Engineering Society 2014; de Kort 2019). Light allows the environmental cognition (mental processing, semantic interpretation and symbolic decoding that involve attention, perception, memory, reasoning, judgment, imagining and thinking) and also stimulate emotions (Tomassoni et al. 2015), thus affecting motivations and influencing purposeful behavioural outcomes.

The urban legibility, "*imageability*" and cognitive understanding (Lynch 1960), along with its appearance and evaluation (Nasar 1994, 1998) are central for human perception of their surroundings. Nasar proposed also a probabilistic model of evaluative response to the image of the city, considering the visual-aesthetic perception and appreciation through the

> favourable emotions and meanings experienced in relation to the environment. (Nasar 1998)

The basic findings of these studies were that human experience of the city is defined by the physical structure of the city, the ease of which the information could be perceived but also by the subjective emotional associations. In the cognitive process, the human perceptual system gathers the meaningful information from the environment and it is influenced by the "*individual's history of exposure to the environment*" (Gibson 1986). Consequently, the perceptual system, and so behaviours, can be considered function of personal (age and gender, memory, personality and sensitivity cultural and social experiences) and environmental factors (tactile, thermal, acoustic and visual perception). Even if highly subjective, aesthetic impressions are similar between individuals with the same sociocultural background and geographical provenience (Nasar 1998).

2.2 The Human and Social Urban Lighting Criteria

The mentioned urban studies focused on the daytime. The same assumptions could be considered true also during night-time but few scientific studies are available to support a theory of the psycho-social influence of lighting on human beings during the night-time of the cities (Kim 2018).

The state of the art of lighting guides, standards, codes of practice and recommendations, aimed at assisting all who take decisions, from lighting designers to municipalities, are mainly focused on safety and security of car drivers and not really committed in considering pedestrian needs with urban lighting. Lighting standards are based on technical elements, and quantifiable, reliable, comparable values. They are aimed at eliminating bad lighting design; on the other hand, good lighting should satisfy also the emotional response of human beings (Boyce 2012).

A brief comparison between lighting codes and guides of practice found that:

- the IESNA handbook (2011) is the most complete guide, providing recommendation for a good practice of urban lighting in relation to pedestrian oriented urban lighting. It focus on the aesthetic impact on visual perception, environmental issues and lighting techniques for different urban elements, differentiating specific areas of the town with design needs and criteria.
- the BS 5489-1: 2020—Design of roads and public amenity areas. Code of practice provides information about the main functions of the urban lighting oriented to pedestrian such as Night-time Identity, Attractive Atmospheric Appearance, Safety and Security, Orientation and Landmarks.
- CIE 115: 2010—Lighting of roads for motor and pedestrian traffic and UNI EN 13201-2 2016 identify specific lighting performance aimed at safety and security of car drivers, by setting a series of street lighting categories based on street typologies. They define quantifiable lighting variables such as illuminance levels, uniformity of illuminance, and glare reduction for pedestrian areas.

From the comparison of these publications, it is evident that they are evolving toward considering more qualitative aspects of lighting. Despite this, they describe the pedestrian-oriented urban lighting in a simplistic way, as a sub-product of street lighting and car-drivers users. Recommendations claim lighting as a positive element in the urban nocturnal environment, an attractor for people to the urban areas and an encouragement tool for social contacts. Despite general statements, few qualitative descriptions are provided without any specific guideline and agreed methodologies for qualitative assessment. Visual comfort, safety and security are mentioned for mixed situations (car drivers and pedestrian) where it is devised a dangerous complexity. Guidelines state that lighting is a critical factor to facilitate obstacle detection and visual orientation, to allow facial recognition that is related to safety perception (Fotios 2013; Institution of Lighting Engineers 2005). More human-oriented or social-oriented features such as accessibility, social fruition and urban appearance are named without guiding further for the lighting design phase.

On the other hand, lighting practitioners claim that urban lighting standards for people should be completely redefined in terms of lighting variables and priorities. As stated by Piero Castiglioni:

> The public space for pedestrian, being a place for socialization, require a particular attention on the illumination on vertical planes (semi-cylindrical illuminance) meanwhile the normative requires horizontal illuminance on the ground and a certain uniformity, another factor that, in my opinion, is not influential and completely counter-productive: we can't intend socialization space like airport service areas. Uniformity is a synonyms of monotony that is not comfortable but a generating factor for anxiety. (Translated transcript excerpt of an interview done by the author to the lighting designer Piero Castiglioni—12/12/2012)

In addition, it is important to achieve a qualitative and quantitative understanding and description of the human and social oriented needs and lighting criteria that concur to qualify human and sociable urban spaces.

> The definition of the terms comfortable, pleasant, relevant and appropriate need no longer be left to the vagaries of "artistry"; they can be defined much more specifically. (Lam 1992)

Currently, there is no general agreement on which are the human and social needs and how they should be either assessed or operationalized. Lighting quality for pedestrian is still an abstract wide concept without an agreement on its definitions. Some researcher (Boyce 2003) stated that it should be defined in terms of its influence on visual performance and behaviours. Veitch and Newsham (1996) instead gave a broader definition, as supporting not only visual performance, task and behavioural performance but also social interactions, mood, health, safety and aesthetic judgements.

The following paragraphs are an attempt to review the studies about the human and social-oriented criteria of lighting for the perception of urban outdoors to evidence five macro-categories that explore the lighting quality for pedestrians: functionality, wellbeing, ambiance and urban beautification, relationship/experience, and environmental issues (Fig. 2.2).

2.2.1 Functionality

2.2.1.1 Accessibility and Spatial Affordance

A space can be defined accessible when it is able to bring people together, guaranteeing the participation of all the members of the community without excluding certain categories. The accessibility of a place can be measured by looking at the physical–visual connections and spatial affordances, defined as the interaction between individual abilities with an environment's spatial organization (Gibson 1986). Physical affordances of spaces are provided with soft edges, shade of transparency from indoor to outdoor that are inviting people, providing interest and containing optional activities (Gehl 2010). Besides, urban lighting should ensure an equal and free access to the entire city to all citizens.

> Lighting should contribute to reducing social and economic inequalities and support integration policies. (LUCI—Charter on Urban Lighting 2007)

2.2.1.2 Visibility, Visual Comfort and Visual Acuity

Functionality is providing a good vision, necessary to comfortably perceive the environment and deriving strictly from the ability to perform visual tasks. In particular, visual acuity is conceived as the capacity for seeing distinctly fine details that have very small angular separation which the observer can just perceive to be separate. Visibility is the range of visual perception, normally measured in terms of the threshold distance at which a sign becomes visible (ILV 2009). Pedestrian lighting is prominently important for ensuring the safe walk along pedestrian paths by providing

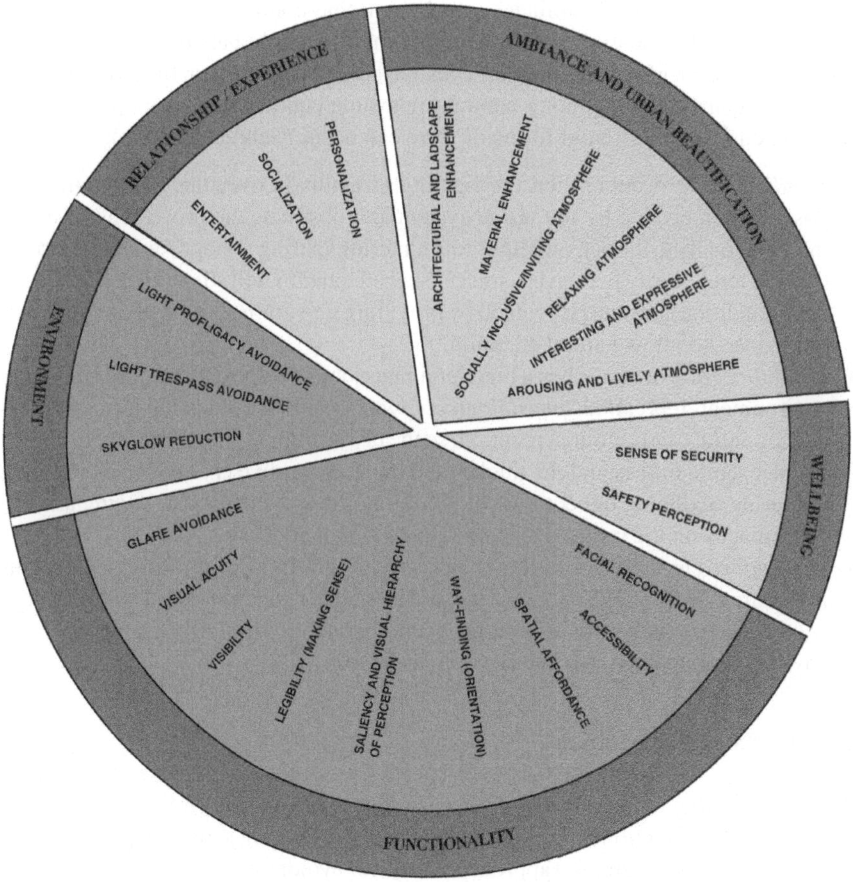

Fig. 2.2 Framework of the human and social urban lighting criteria

clear and detailed vision to understand the properties of the surfaces, to avoid obstacles and people, and to prevent the risk of falls and collisions. Both visibility and visual acuity are related to the concept of visual comfort that, in outdoor environment is achieved thanks to glare avoidance in all the conditions of vision and an appropriate lighting distribution in the space (Hargroves 2001). The human visual system is subject to natural aging resulting in decreased visual capabilities that can reduce visual comfort in the outdoor environment at night. Elderly people suffer for reduced visual acuity, resulting in limited ability to detect obstacles. Hence, they need higher lighting levels.

2.2.1.3 Glare

Glare defines the discomfort and the reduction in the ability to see caused by an unsuitable distribution of luminance, or by extreme contrasts in the field of view. It is influenced by the eye's adaptive ability and the angle between the task to see and the source of glare. The human eyes have not evolved over time to deal with night-time glare since it is not a natural night-time condition. Considering outdoor environments, three different forms of glare can occur (Schreuder 2008):

- *Blinding glare* occurs when the lighting stimulus is over the maximum value possibly perceivable by the visual system and obstructs the processing of useful information from the surroundings (e.g. entering/exiting a shop from/to the street). Glare is temporarily perceived, since visual adaptation will adjust to the new luminous condition, and can be avoided by providing transition zones with intermediate luminance, to allow natural adaption.
- *Disability glare* occurs when a high luminance object is located in a low luminance environment. It is called "*physiological glare*": vision is reduced in terms of luminance contrast and colour perception, so the task is impossible to be seen. In elderly people, senile miosis and the thicker and more absorptive crystalline increase the scattering, reducing the contrast of the retinal image, creating a "*luminous veil*" over images on the retina (Dejeammes and Chain 2010).
- *Discomfort glare* occurs in the presence of bright light of lighting sources and luminaires in the field of view. It is often called "*psychological glare*" and it is considered much more important in comparison to disability glare in outdoor environment, mainly in relation to car drivers.

2.2.1.4 Facial Recognition

It is the possibility to identify faces' expressiveness at a distance for enabling people to anticipate the intentions of approaching figures by looking at their expressions and to react in time accordingly (approachability vs. avoidance) (Fotios and Raynham 2011). The concept of facial recognition is based on the environmental studies developed by Hall (1966) that defined four different personal zones around an individual: intimate, personal, social-consultative and public. From his study, people were found to feel uncomfortable when strangers entered their personal spaces (at distances of less than 3 m) unless they could recognize them. Personal spaces can change according to personality, culture and environmental factors, such as low illumination (Adams and Zuckerman 1991). Van Bommel and Caminada (1982) correlated facial recognition to the increase of feelings of safety and introduced the semi-cylindrical illuminance (E_{SC}) measured at of 1.5 m. It describes the amount of vertical illuminance (E_v) that falls on a semi-cylindrical surface (i.e. a hypothetical human face) to recognize expressions. Distance and eyes adaptation are also other factors in people recognition of expressions. According to Rombauts et al. (1988), an E_{SC} of 0.6 lx, (roughly equivalent to a value of horizontal illuminance (E_H) of 5 lx is the minimum value that let people recognize faces at a distance of 4 m (sufficient for evasive or defensive actions). At 10 m distance, which would give greater time for any necessary avoiding action the recommended level is 2.7 lx (CIE 136 2000). Based on several further studies (Boyce 2003), lighting levels of E_{SC} were identified for different pedestrian street categories and they were also correlated to E_V and E_H, thus defining pedestrian lighting classes in the normative. More than this, the influence of spectral power distribution (SPD) on facial recognition resulted controversial: some studies (Fotios and Cheal 2011) found that higher colour rendering is better than low due to the perceived spatial brightness of different SPD at the same illuminance levels. Other studies (Yang and Fotios 2014) found no effects of SPD on facial recognition.

Other than face, people could also observe behaviours and postural attitudes to approach or avoid other person, but the studies about this aspects are very few in the literature.

2.2.1.5 Urban Legibility

The urban legibility is related to the primitive need to understand the environment to extract the necessary information for survival (Rapoport 1977). It refers to the ease and clarity with which the different parts of the city can be recognized and organized by its inhabitants (Lynch 1960).

> Understanding provides a sense of security. When people cannot understand a situation, they can become distressed. (Kaplan et al. 1998)

A legible city allows for an easy identification of its main elements and for efficient navigation and wayfinding. Lynch's work of cognitive maps to orientate and navigate the space has made a constant reference to light as a prominent form shaper, to increase the distinctive quality to space. In this regard, lighting's primary aim is to reveal the environment, to orient people gaze by shaping the physical surroundings and thresholds. It is the instrument enabling to see position and direction but also to make inferences about distances and depth.

However, his theories has been tested during the day when all elements of the city are fully visible. Besides, at night, only few elements can be lit by artificial means, providing a different appearance than the one defined by the sun and the sky. Lighting could influence the urban nocturnal legibility, by visually extending or limiting the space, guiding the eyes across the space due to the luminous contrasts between objects, emphasizing some urban elements and blanking others. It can reveal details and the structure and volumes to better understand the space at night. The more the space is unknown to people, the more is important to provide lighting for guiding the eyes in providing directions and making people reading the space safely without any misleading information.

2.2.1.6 Spatial Recognition, Visual Orientation and Way-Finding

Spatial recognition, orientation and way-finding are also extremely functional in perceiving the space. With the aid of artificial lighting, it means to create night-time focal points that can help also to measure the distances and depth by the reciprocal position of objects, to support the extraction of necessary information for navigation. The use of visual luminous contrast can help the understanding and legibility of the space. Visual contrast is defined as the perception of a difference of light (luminance) coming from two adjacent areas between the target and its background. Visual contrast should be used properly in the urban environment with orienting and informing purpose: lighting distribution can codify messages, support information and simplify the reading of the space with a cognitive, aesthetic and symbolic rule (Kepes 1966).

Way-finding and depth perception can be enhanced by the proper use of lighting distribution, lighting intensity variation, changes in colour temperature of white and also in colour of lighting. Orientation in different areas of the city can be enhanced by different lighting techniques such as

> variation in lighting pattern, colour and illuminance from one area to the next one. (The Outdoor Environment 1992)

in order to define the whole nightscape but also to underline the differences between districts within the city area maintaining a visual coherency and composing hierarchy.

Different lighting techniques enhance spatial recognition and navigation: accent bright lighting among specific urban elements, architecture and paths may attract visual attention and is invariably more effective than uniform high levels of lighting (Moyer 1992). He found that the control of brightness distribution can direct spatial perception and people movements in relation to where their attention is directed. Attention and interest of the viewer are drawn to brightly lit objects even if they are located in the background, in comparison to softly lit objects located in the foreground. High contrasts between two contiguous areas confound the understanding of the space because the eyes start to "bounce" among the twos with an unpleasant effect. A lower lighting level that fills the spots is considered more comfortable.

Other studies examined the influence of the luminance distribution, intensity and placement of lighting on attention guidance and spatial behaviours in interior contexts such as classrooms to enhance the attention of students toward specific illuminated tasks (La Giusa and Perney 1974), in corridor selection to drive people movements through the brighter one (Taylor and Socov 1974), in a restaurant setting to orient and select place (Flynn et al. 1973), in museums to select circulation routes and to linger in the space by interacting with the setting (Antonakaki 2006). In outdoor situations, people driving car were found to be attracted by lighting when they oriented their eyes toward source of light at the corner of their visual field. Despite this, the main activity of driving was not evidently disturbed or any change in direction occurred (Green 2006). This phenomena can be explained by the fact that the sight at night is driven by luminous points and luminous spaces: the "*moth effect*" is expressed by gaze attraction driving people exploration and perception of the space.

2.2.1.7 Saliency and Visual Hierarchy of Perception

Saliency and visual hierarchy are related to the previous point and deal with the property of producing features' contrast through appropriate brightness distribution of the scene to suit individual efficient visual search of urban objects or urban activities, to establish a visual hierarchy, to draw attention to landmarks. Existing guidelines recommend higher luminance contrast between urban object and background for higher saliency (or degree of conspicuity) (ILE 2005). Besides, there are also other factors affecting the visual saliency: luminance contrast between the target and the background and the density of lighting patterns in the urban background in relation to the target. The saliency of urban objects can decrease if there is not enough contrast of luminance against its background or if the background is very dense of light patterns (Davoudian 2011). Another study (de Sousa Del-Negro 2015) demonstrated also that lighting could become a distractor, by distorting the image, size, dimensions and volumetric image of landmark that may be detected but not recognised when lit with artificial lighting.

2.2.2 *Wellbeing*

The sense of wellbeing is interlinked with comfort, either visual or psychological in experiencing the nocturnal urban space. Night-time streets can be of discomfort to pedestrian when hiding details and source of danger and this causing a sense of insecurity. Accordingly, public lighting was originally introduced into urban areas as a mean for crime and vandalism reduction both toward people and properties (Raynham and Saksvikrønning 2003). Wellbeing includes safety, security, safety perception and reassurance.

2.2.2.1 Security

The sense of security is related to personal security from traffic or other hazards, in mixed situations with pedestrian, or where it is devised a dangerous complexity, and also in urban areas where the crime risk is higher than normal. The visual tasks involved are quite simple and associated with locating individuals and avoiding obstacles on the footpath. Higher level of illuminance, uniformly distributed without glare are adopted to provide better visibility of the surroundings and contrast between the pedestrian and the background (Boyce 2003; Van Santen 2006; Moyer 1992). More than this, elderly and visually impaired people need higher illuminance level and better contrast (avoiding glare), to provide good visibility and security in the nocturnal outdoor walking activities.

2.2.2.2 Safety Perception

The correlation of safety and lighting is highly debated: if certain studies (Painter 1988; Tien 1979) stated that improved lighting results in less outdoor criminal behaviours, due to the increase of visibility and the increase of social control of the community. Conversely, other studies (e.g. Farrington and Welsh 2002) found the opposite outcomes. Besides, the research gradually shifted from the measured safety (evaluated in terms of decreasing number of illegal and criminal acts) to the perception of safety (that is a more subtle and subjective impression). From counting the quantity of committed crime, studies passed on evaluating the fear of crime and the perceived risk of being a victim of crime. Many studies (Welsh and Farrington 2008; Pease 1999) have explored the issue of the relationship between lighting and safety, concluding that safety perception is dependent on a series of factors other than solely outdoor lighting. Perceived safety is influenced by personal factors such as personality and attitudes and by environmental factors such as the perception of spatial brightness and the hedonic tone of the space (pleasantness, naturalness, interest) (Boyce 2003). Spatial brightness implies better visual performance to see finer details at greater distances (Boyce 2003), hence enable people to read the surroundings and the facial and body language of approaching people.

Fear of crime is related to darkness or shadows perception and safety is associated with bright and uniform illuminated streets (Nasar and Jones 1997; Hanyu 1997).The perception of safety of parking lots and street sidewalk was found in non-linear correlation with increasing E_H. For illuminances comprised between 0 and 10 lx, small increases in illuminance produced large increases in perceived safety. For illuminances above 50 lx, increase in illuminance made little difference to perceived safety. In the range of 10–50 lx, increases in illuminance show a law of diminishing returns. These results suggest that in urban and suburban area, an horizontal illuminance around 30 lx is required for good perceived safety (Boyce et al. 2000). In addition to lighting levels, other qualitative lighting variables contributes to the perception of safety in the outdoor environment. The lighting distribution, static or dynamic, the SPD, the direction and height of urban lighting fixture which concur to glare effects, can affect the feelings of safety. About dynamic lighting in relation with lighting distribution, Haans and De Kort (2012) investigated which areas of the road pedestrians prefer to have illuminated with respect to their perceived personal safety. Stationary and walking pedestrians were found to prefer having light in their own immediate surroundings in comparison to lighting more distant parts of the road.

2.2.2.3 Reassurance

Pedestrian reassurance could be described as the level of confidence that people have in walking a street alone after dark street. It is related to the fear of crime and perception of safety, either real or imaginary (Unwin 2014). Studies (Boyce et al. 2000; Vrij and Winkel 1991) suggest that higher illumination may contribute to pedestrian reassurance and could encourage the overall social use of the city. Further research (Fotios et al. 2014) on uniform lighting distribution and lighting levels has been claimed as important to understand the proper lighting strategies acting for the *situational crime prevention model* (Welsh and Farrington 2008). Besides, the presence of other people is important to guarantee the psychological comfort for people and it can be an effective tool in promoting outdoor safety, thus reassurance. When people are in the space, the increase of safety perception is based on the community pride, informal social control, natural surveillance, and sense of community. This method for crime prevention is based on the *strengthening informal social control* (Welsh and Farrington 2008).

2.2.3 Ambiance and Urban Beautification

Lighting can increase the urban nocturnal atmosphere for a more interesting and appealing impression of places, thus motivating in visiting or spending time in urban outdoor (Peters 1992).

2.2.3.1 Architectural, Material and Landscape Enhancement

Outdoor lighting shapes the urban character, image and identity (Peters 1992). It can enhance the architectural volumes and the landscape composition, highlighting details and defining hierarchies to boost the attractiveness of the city. Lighting can increase the city ambiance, that is defined as the aesthetics and the attractiveness of the city and subsequently the comfort of people within it. Urban lighting can provide a general feeling of well-being contributing actively to "beautification", "pleasantness" e "amenity" (Hargroves 2001) of the city.

The overall appearance of public areas is due to night-time appearance and pedestrian comfort that are dependent to features of the lighting sources (e.g. lighting colour appearance and rendering), features of the lighting fixtures (e.g. luminous appearance of the luminaire and glare), features of the lighting installation (e.g. lighting spatial distribution and lighting dynamics both in terms of perceived luminous contrast ratio and lighting behaviours) and features derived by the relationship between lighting and the environment (e.g. brightness perception, material reflectance, illuminance and urban colour appearance). The environmental characteristics of materials and the structure of the city can be highlighted through lighting that increase the perceivable environmental characteristics of the urban matters such as colours, textures and patterns.

2.2.3.2 Atmospheres

The evocativeness of the luminous atmosphere in the urban space concur in creating different impressions of the nocturnal urban environment and different emotional responses, even in a familiar space, through the organization of lighting elements (Moyer 1992). Different social situations and environmental contexts require different lighting conditions that have a similar trend of subjective preferences (Butler and Biner 1987; Biner et al. 1989; Kobayashi et al. 2001). Lighting can determine (i) socially inclusive/inviting, (ii) relaxing, (iii) interesting and expressive, (iv) arousing and lively luminous atmospheres.

Studies, mainly in interiors, claim that lighting could create socially inclusive/inviting atmosphere (i) supporting social interpersonal judgements and positive social behaviours such as friendliness, helpfulness, trustworthiness (Baron et al. 1992; Kombeiz 2017; Heijboer 2013). Lower direct lighting levels with bright illumination at a distance send socially inclusive "*patterns of impressions*" to which individuals react in consistent ways (Flynn et al. 1973; Flynn 1988). Dim, dark and not uniform lighting conditions are chosen for private/intimate social situations requiring low behavioural control (e.g. relaxing, listening to music, thinking, taking a break, talking with a friend, dining with the partner, having a party, being romantic and love making) in contexts such as home (living room), office, classroom and library (Butler and Biner 1987; Biner et al. 1989; Kobayashi et al. 2001).

Lighting could support a comfortable, calming and hospitable experience (ii) (warm and non-uniform lighting distribution) (Flynn and Spencer 1977).

Interesting and expressive atmosphere (iii) can be achieved using a combination of the techniques coined by Kelly (1952), such as *focal glow, ambient luminescence and play of brilliants*. They are three categories of lighting distribution to articulate the environment starting from the human visual perception of scenes. *Ambient luminescence* refers to uniform general lighting that enables the observer to get their bearing and provides a feeling of safety and "*minimizes the importance of all things and people*" (Kelly 1952). It is necessary for revealing the night form of the city creating a comfortable restful and reassuring place. *Focal Glow* or accent lighting is used to emphasize objects, areas and specific zones to create hierarchies of perception to command attention, attracting interest and pointing the important element of the night cityscapes. It was described as the "*campfire of all time*" feeling the sensation of proximity and warmth of lighting. *Play of brilliant* is the decorative light that results in surprise and amazement through dynamic colour progressions or light patterns projections used to create visual attraction. It is encapsulated in the idea of "*Times Square at night*". Night city scenery is more appreciated and appealing when combining different lighting effects: people were found to prefer lighting schemes that provide more contrast and shading, with a layered approach. This means a good general uniform illuminance coupled with visual accents and decorative lighting elements that provide the highlight of certain features within the environment such as trees, seating, entrances (Hargroves 2001).

Lighting could determine positive arousal, pleasure and excitement (iv) due to the novelty, originality and unexpected or unconventional surprising ways of using light for experiencing the night-time. Studies have shown that human beings require a balance of unity and complexity in the built-environment because the natural conditions they are used to are continuously changing (Birren 1983). People expect all senses to be moderately stimulated in the built-environment: the lack of complexity results in adverse psychological reactions (Birren 1969). A balanced constantly changing environment can determine a positive, aroused perception. In lighting design, Mahnke (1996) stresses the importance of providing a good balance of variety and unity: variety is necessary for arousal and unity for a favourable impression in order to avoid sensory fatigue (Kaplan and Kaplan 1989).

2.2.4 *Relationship/Experience*

2.2.4.1 Personalization, Socialization and Interaction

Urban lighting could have an agent purpose, making the nocturnal urban environment suitable for its social passive or active use, adaptable to context and people presence. This is also due to the technological upgrade of technologies toward the *Ledification* and *Digitization* of outdoor urban lighting. Nowadays, there is a transformation of lighting design in the cities due to the development of controllable and networkable LED lights that could redefine the static lighting into dynamic and intelligent luminous performance and schemes.

At the simplest level, by increasing the urban ambiance, therefore the perceived feeling of comfort, safety and pleasantness of the city at night can increase pedestrian activity (Painter 1994), and so sudden social passive or active encounters (Jacobs 1995) and *socialization.* A small amount of studies (mainly in interiors) has explored the effects of lighting on spatial behaviours on a personal space and social level. Dim lighting conditions enhanced prosocial behaviours (Werth et al. 2012; Steidle et al. 2013), promoted interpersonal closeness (Kombeiz et al. 2017) reduced social distance to achieve affiliation, disclosed more information in intimacy situation (Carr and Dabbs 1974; Miwa and Hanyu 2006; Evensen 2014). Conversely, dim lighting can make the not-intimates feel more uncomfortable and all the consequent nonverbal behaviours appear to be compensatory efforts to re-establish more psychological distance. Kobayashi (2013) also found that lighting can improve interpersonal relationship and appropriate communication such as more sociopetal orientation and body angling between people during the night (leisure time), with differences occurring between gender and culture. Even though, judged privacy increased with non-uniform, dim, and peripheral lighting in a study conducted on urban squares

(Nasar and Bokharaei 2016), people rated more favourable the uniform and bright lighting conditions in public spaces after dark.

Human-size lighting performances provide a human connection with the space through the possible *personalization* of the luminous atmosphere of the urban environment and also through the possible activation of people engagement in playful, unconventional *interaction.*

The on-time tangible lighting encounter with people in urban night-time is delivered through a simple physical participation (detection of presence and movements) and determines a more natural and subtle interaction in the routinely use of the urban environment that can be seen or ignored. This also define a very strong relationship and a sense of personalization of the luminous environment (Casciani and Rossi 2015). Embodied and physical ways of interaction with lighting fixtures and lighting effects are more preferred rather than virtual and digital ones (Poulsen et al. 2013a, b). Bodily interaction with lighting increase the level of personal contact, control and transformation of lighting and city atmosphere. Some studies (Poulsen et al. 2012, 2013a, b) explored the effects of urban lighting with implicit interaction (tracking of position and movements of people), with explicit interaction (adaptation to wind patterns), as well as participation, influencing people to stop and interact in a playful manner with lighting and, as a consequence, between users. The provision of noticeable lighting levels in the near proximity of the users and the lighting adaptation in real time following people direction and movements was perceived as contributing more to safety perception and comfort with a pleasant feeling of personalization (attachment) (Haans and De Kort 2012). Thomsen (2009) also studied the social dimension of interactive street lighting designing and testing lighting effects (increasing level or changing colour according to speed and quantity of people) to stimulate playful behaviours. In this cases, people were actively participating in the making of the lighting performance both as actors, makers and players meanwhile a new culture of public space was formed. The entertaining performance enact social exchange, communication and playfulness, by actively involving people in participating. The transformation of lighting in a subtle and nonobtrusive way in relation to presence and behaviours can favour people retrieved sense of personal control over the lighting performances. It also can adapt to urban patterns and rhythms, create new aesthetic experience, convey social meaning and also contribute to energy efficiency (Pihlajaniemi et al. 2013). Finally, proper playful social interaction with lighting has to be based on the study on people and context of use (Palmer and Popat 2008):

> interaction design needs to focus on site-specific technology, context-aware systems and location-based services that create enduring cultural frameworks. (McCullough 2004)

2.2.5 *Environmental Issues*

2.2.5.1 Skyglow Reduction, Light Profligacy Avoidance and Light Trespass Avoidance

Night light pollution is problematic due to its impacts on the health of human and of the ecosystem, disrupting the natural ecological processes such as birds migration and orientation, animal reproduction, foraging and predation, and affecting human circadian rhythms (Navara and Nelson 2007). In addition to this, the overuse of night artificial light is reducing visibility of the starry sky and finally wasting energy (Lighting Research Center 2003).

The phenomena of lighting pollution takes different forms. Here we discuss Skyglow, Light Profligacy and Light Trespass. Achieved through the reduced combination of reflected and refracted light from the atmosphere, skyglow reduction is aimed at limiting contrast in the sky. It is determined by the cumulative effect of the illumination of lighting bouncing off of microscopic particles suspended in the air. If light profligacy avoidance is defined as the limitation of over-illumination, light trespass avoidance is the limitation of the unwanted light that is spilling into private properties and activities (e.g. light that extends onto the facade of a building, into windows, or into the sky).

Phenomena such as over-illumination (the more light the best), continual increase of the population, the exponentially growing energy depletion are claiming for a different way of designing urban lighting. Eco-responsible practices focus on the energy consumption reduction through the active responsibility of users, even in outdoor urban places.

> [...] the first question that anyone concerned with purchasing or designing a new outdoor lighting installation should ask themselves is 'Is this lighting necessary and, if it is, when is it necessary? (Boyce 2003)

The responsive and adaptable lighting, coherently matched with the activities of users, could support the twist toward more environmental choices.

References

Adams L, Zuckerman D (1991) The effect of lighting conditions on personal space requirements. J Gen Psychol 118(4):335–340. https://doi.org/10.1080/00221309.1991.9917794

Antonakaki T (2006) Lighting within the social dimension of space: a case study at the Royal Festival Hall. In: Proceedings of the workshop held in Bremen

Baron R, Rea M, Daniels S (1992) Effects of indoor lighting (illuminance and spectral distribution) on the performance of cognitive tasks and interpersonal behaviors: the potential mediating role of positive affect. Motiv Emot 16(1):1–33. https://doi.org/10.1007/bf00996485

Bell P, Greene T, Fisher J, Baum A (2001) Environmental psychology, 5th edn. Psychology Press, Taylor & Francis Group, New York

Biner P, Butler D, Fischer A, Westergren A (1989) An arousal optimization model of lighting level preferences. Environ Behav 21(1):3–16. https://doi.org/10.1177/0013916589211001

Birren F (1969) Physiological implication of colour and illumination, the psychology of colour and illumination. J Illum Eng
Birren F (1983) Color and psychotherapy. Interior Design 54(12):166–169
Boyce PR (2003) Human factors in lighting, 2nd edn. CRC Press, London
Boyce PR (2012) Editorial: too much information? Lighting Res Technol 44(3):259–259. https://doi.org/10.1177/1477153512456993
Boyce PR, Eklund N, Hamilton B, Bruno L (2000) Perceptions of safety at night in different lighting conditions. Lighting Res Technol 32(2):79–91. https://doi.org/10.1177/096032710003200205
BS 5489-1 (2020) Design of roads and public amenity areas. Code of practice
Butler D, Biner P (1987) Preferred lighting levels. Environ Behav 19(6):695–721. https://doi.org/10.1177/0013916587196003
Carr S, Dabbs J (1974) The effects of lighting, distance and intimacy of topic on verbal and visual behavior. Sociometry 37(4):592. https://doi.org/10.2307/2786430
Casciani D, Rossi M (2015) Exploring the social dimension of urban lighting. In: 2nd Iran lighting design conference (ILDC) proceeding, Teheran
CIE 115 (2010) Lighting of roads for motor and pedestrian traffic
CIE 136 (2000) Guide to the lighting of Urban Areas
Cuttle C (2008) Lighting by design. Butterworth-Heinemann/Elsevier, Amsterdam
Davoudian N (2011) Visual saliency of urban objects at night: impact of the density of background light patterns. LEUKOS 8(2):137–152. https://doi.org/10.1582/leukos.2011.08.02.004
de Sousa Del-Negro D (2015) The image of the city at night: the effects of artificial lighting on urban legibility and wayfinding. PhD, UCL Institute for Environmental Design and Engineering, Faculty of the Built Environment
Dejeammes M, Chain C (2010) Accessible street environment for people with vision impairment: research for new specification sin France. In: Transed 2010: 12th International conference on mobility and transport for elderly and disabled persons, 1–4 June 2010
Evensen I (2014) A study on the effects of lighting on social interaction (Undergraduate). The Bartlett School of Graduate Studies
Farrington D, Welsh B (2002) Effects of improved street lighting on crime: a systematic review. Dev Stat Direct 19(2):313–342. https://doi.org/10.1080/07418820200095261
Flynn J (1988) Lighting design decisions as interventions in human visual space. In: Nasar J (ed) Environmental aesthetics. Theories, research, & applications. Cambridge University Press, Cambridge
Flynn J, Spencer T (1977) The effects of light source color on user impression and satisfaction. J Illum Eng Soc 6(3):167–179. https://doi.org/10.1080/00994480.1977.10747811
Flynn J, Spencer T, Martyniuk O, Hendrick C (1973) Interim study of procedures for investigating the effect of light on impression and behavior. J Illum Eng Soc 3(1):87–94. https://doi.org/10.1080/00994480.1973.10732231
Fotios S (2013) LRT digest 1. Maintaining brightness while saving energy in residential roads. Lighting Res Technol 45:7–21. https://doi.org/10.1177/1477153512464141
Fotios S, Cheal C (2011) Predicting lamp spectrum effects at mesopic levels. Part 1: Spatial brightness. Lighting Res Technol 43(2):143–157. https://doi.org/10.1177/1477153510393932
Fotios S, Raynham P (2011) Correspondence: lighting for pedestrians: is facial recognition what matters? Lighting Res Technol 43(1):129–130. https://doi.org/10.1177/1477153511400158
Fotios S, Unwin J, Farrall S (2014) Road lighting and pedestrian reassurance after dark: a review. Lighting Res Technol 47(4):449
Gehl J (2010) Cities for people. Island Press, Washington, D.C.
Gibson JJ (1986) The ecological approach to visual perception. LEA, London
Green M (2006) Is the moth effect real? Accid Reconstr J 16(3):18–19
Haans A, de Kort Y (2012) Light distribution in dynamic street lighting: two experimental studies on its effects on perceived safety, prospect, concealment, and escape. J Environ Psychol 32(4):342–352. https://doi.org/10.1016/j.jenvp.2012.05.006
Hall ET (1966) The hidden dimension. Anchor Books, New York

Hanyu K (1997) Visual properties and affective appraisals in residential areas after dark. J Environ Psychol 17(4):301–315. https://doi.org/10.1006/jevp.1997.0067
Hargroves R (2001) Lighting for pleasantness outdoors. In: Proceedings of the CIBSE national conference. Chartered Institution of Building Services Engineers, London
Heijboer M (2013) Seeing things in a positive light. Investigating the effects of color temperature of light on sociality and de-escalation. master of science in human technology interaction, Department of Industrial Engineering & Innovation Sciences, Eindhoven University of Technology
IES handbook, 10th edn (2011)
ILE—Institute for Lighting Engineers (2005) The outdoor lighting guide. Taylor & Francis, New York
ILV: International Lighting Vocabulary (2009) CIE DS 017.2/E:2009
Jacobs AB (1995) Great streets. MIT Press, Cambridge
Kaplan R, Kaplan S (1989) The experience of nature: a psychological perspective. Cambridge University Press, Cambridge
Kaplan R, Kaplan S, Ryan RL (1998) With people in mind: design and management of everyday nature. Island Press, Washington, DC
Kelly R (1952) Lighting as an integral part of architecture. College Art J 12(1):24. https://doi.org/10.2307/773361
Kepes G (1966) Light as a creative medium. Arts and Architecture
Kim D (2018) Development of a psychological pathway model linking lighting quality to well-being in indoor café environments. Indoor Built Environ 27(3):390–401. https://doi.org/10.1177/1420326x17753512
Kobayashi S (2013) Lighting effects on unconscious human behaviors. In: 10th International symposium on sustainable healthy buildings annual spring conference. Society of Lighting & Visual Environment, Seoul
Kobayashi S, Inui M, Nakamura Y (2001) Preferred illuminance non-uniformity of interior ambient lighting. J Light Visual Environ 25(2):64–75. https://doi.org/10.2150/jlve.25.2_64
Kombeiz O (2017) Turning the spotlight on the role of light and colors in offices: how are performance, social interactions, and social perception affected? PhD, University of Hohenheim
Kombeiz O, Steidle A, Dietl E (2017) View it in a different light: mediated and moderated effects of dim warm light on collaborative conflict resolution. J Environ Psychol 51:270–283. https://doi.org/10.1016/j.jenvp.2017.04.007
de Kort YAW (2019) Tutorial: theoretical considerations when planning research on human factors in lighting. LEUKOS 1–12. https://doi.org/10.1080/15502724.2018.1558065
de Kort YAW, Veitch JA (2014) From blind spot into the spotlight. J Environ Psychol 39:1–4. https://doi.org/10.1016/j.jenvp.2014.06.005
La Giusa FFL, Perney LR (1974) Further studies on the effects of brightness variations on attention span in a learning environment. J Illum Eng Soc 3(3):249–252
Lam W (1992) Perception and lighting as formgivers for architecture, 1st edn. Van Nostrand Reinhold, New York
Light + Behavior Symposium – Illuminating Engineering Society (2014) Ies Org. https://www.ies.org/research/events/research-symposia/light-behavior-symposium/
Lighting Research Center (2003) Implementation of decision-making tools that address light pollution for localities planning street lighting. Whyte paper efficient street lighting and light pollution
LUCI—Charter on Urban Lighting (2007)
Lynch K (1960) The image of the city. The MIT Press, Cambridge
Mahnke FH (1996) Color, environment, and human response: an interdisciplinary understanding of color and its use as a beneficial element in the design of the architectural environment. Wiley, Hoboken
McCullough M (2004) Digital ground: architecture, pervasive computing and environmental knowing. MIT Press, Cambridge

Miwa Y, Hanyu K (2006) The effects of interior design on communication and impressions of a counselor in a counseling room. Environ Behav 38(4):484–502. https://doi.org/10.1177/0013916505280084

Moyer JL (1992) The landscape lighting book. Willey, New York

Nasar J (1994) Urban design aesthetics: the evaluative quality of building exteriors. Environ Behav 26(3):377–401. https://doi.org/10.1177/001391659402600305

Nasar JL (1998) The evaluative image of the city. Sage, Thousand Oaks

Nasar J, Bokharaei S (2016) Impressions of lighting in public squares after dark. Environ Behav 49(3):227–254. https://doi.org/10.1177/0013916515626546

Nasar JL, Jones KM (1997) Landscapes of fear and stress. Environ Behav 29(3):291–323. https://doi.org/10.1177/001391659702900301

Navara KJ, Nelson RJ (2007) The dark side of light at night: physiological, epidemiological, and ecological consequences. J Pineal Res 43(3):215–224. https://doi.org/10.1111/j.1600-079x.2007.00473.x

Painter K (1988) Lighting and crime prevention: the Edmonton project. Middlesex Polytechnic, London

Painter K (1994) The impact of street lighting on crime, fear of crime and pedestrian activity. Secur J 5:116–124

Palmer S, Popat S (2008) FEATURE dancing in the streets. Interactions 15(3):55. https://doi.org/10.1145/1353782.1353797

Pease K (1999) A review of street lighting evaluations: crime reduction effects. In: Painter K, Tilley N (eds) Surveillance of public space: CCTV, street lighting and crime prevention. Crime prevention studies, vol 10. Criminal Justice Press

Peters RC (1992) The language of light [Light in Place]. J Issue Places 8(2). (Places, College of Environmental Design, UC Berkeley)

Pihlajaniemi H, Österlund T, Herneoja A (2013) Towards adaptive urban lighting; algorithm aided design of experiential urban lighting. In: Arkkitehtuuritutkimuksen päivät 2011: Tutkimus ja käytäntö (Proceedings of the 3rd symposium of architectural research 2011: research & praxis) Arkkitehtuuritutkimuksen päivät 2011, Oulu, pp 164–180

Poulsen ES, Andersen HJ, Jensen OB, Gade R, Thyrrestrup T, Moeslund TB (2012) Controlling urban lighting by human motion patterns results from a full scale experiment. In: Proceedings of the 20th ACM international conference on multimedia. ACM, pp 339–348

Poulsen ES, Morrison A, Andersen HJ, Jensen OB (2013a) Responsive lighting: "The city becomes alive". In: MobileHCI'13, 27–30 Aug 2013, Munich, Germany

Poulsen ES, Morrison A, Andersen HJ, Jensen OB (2013b) Responsive lighting: the city becomes alive. In: Proceedings of the 15th international conference on human-computer interaction with mobile devices and services. ACM, pp 217–226

Rapoport A (1977) Human aspects of urban form: towards a man-environment approach to urban form and design. Pergamon Press, Oxford

Raynham P, Saksvikrønning T (2003) White light and facial recognition. Lighting J 68:1

Rombauts P, Vandewyngaerde H, Maggetto G (1988) Minimum semicylindrical illuminance and modelling in residential area lighting. Light Res & Technol 21(2):49–55

Schreuder D (2008) Outdoor lighting: physics, vision and perception. Springer Science, Berlin

Steidle A, Werth L (2013) Freedom from constraints: darkness and dim illumination promote creativity. J Environ Psychol 35:67–80. https://doi.org/10.1016/j.jenvp.2013.05.003

Steidle A, Hanke E, Werth L (2013) In the dark we cooperate: the situated nature of procedural embodiment. Soc Cogn 31(2):275–300. https://doi.org/10.1521/soco.2013.31.2.275

Taylor L, Socov E (1974) The movement of people toward lights. J Illum Eng Soc 3(3):237–241. https://doi.org/10.1080/00994480.1974.10732257

The Outdoor Environment (1992) CIBSE lighting guide LG6. The Chartered Institution of Building Services Engineers, London

The SLL Lighting Handbook (2009) The society of light and lighting printed in England

Thomsen BS (2009) Performative environments. Doctoral dissertation, Aalborg University, The Faculty of Engineering and Science, Urban Design
Tien JM (1979) Street lighting projects. Department of Justice, Law Enforcement Assistance Administration, National Institute of Law Enforcement and Criminal Justice, Washington, DC
Tomassoni R, Galetta G, Treglia E (2015) Psychology of light: How light influences the health and psyche. Psychology 06(10):1216–1222. https://doi.org/10.4236/psych.2015.610119
UNI EN 13201-2 (2016)
Unwin J (2014) An investigation of the effect of street lighting. PhD, The University of Sheffield
Van Bommel WJM, Caminada E (1982) Considerations of the lighting of residential areas for non-motorised traffic. In: CIBSE national lighting conference
Van Santen C (2006) Light zone city: light planning in the urban context. Walter de Gruyter, Berlin
Veitch JA, Newsham GR (1996) Determinants of lighting quality II: research and recommendations. In: Annual meeting of the American psychological association, Toronto, Canada
Vrij A, Winkel F (1991) Characteristics of the built environment and fear of crime: a research note on interventions in unsafe locations. Deviant Behav 12(2):203–215. https://doi.org/10.1080/01639625.1991.9967873
Welsh B, Farrington D (2008) Effects of improved street lighting on crime. Campbell Syst Rev 4(1):1–51. https://doi.org/10.4073/csr.2008.13
Werth L, Steidle A, Hanke E (2012) Getting close in the dark: darkness increases cooperation. In: Experiencing light. International conference on the effects of light on wellbeing, Eindhoven
Yang B, Fotios S (2014) Lighting and recognition of emotion conveyed by facial expressions. Lighting Res Technol 47(8):964–975. https://doi.org/10.1177/1477153514547753

Chapter 3
The Assessment of Affective, Social Appraisal and Behaviours of Human Scale Lighting Experience

Abstract Several studies have been exploring the relationship between lighting variables, urban environment and people perception to understand their reciprocal influence. Through a literature review of indoor and outdoor lighting research, the chapter focuses on understanding the modalities and location of experiential fruition (in-situ real location, laboratory installation, image- and visualization-based experiences, and immersive experiences through augmented reality technologies). The research has also evidenced the different research procedures, methods and tools to develop technical/objective and subjective environmental assessment, highlighting the specific features related to data acquisition (direct elicitation and observation techniques) and data elaboration. The aim is to summarise traditional and experimental methodologies, to report advantages and disadvantages, to evidence the importance of exploring in a scientifically reliable way the affective, social appraisal and behaviours of laypersons in relation to the lighting design and luminous atmospheres of the nocturnal urban environment. Besides, the implication in practice has been discussed highlighting the importance of the systematic use of objective and subjective assessment of outdoor lighting before, during and after a project of lighting design to educate lighting designers and decision-makers toward more human-oriented lighting practice.

Keywords Subjective environmental assessment · Objective environmental assessment · Research procedures · Research methods · Research tools

3.1 Assessing the Experience of Lighting at Night

Few studies about lighting design for interior spaces have already clearly established the effects of lighting parameters on affective, spatial and social appraisal of human scale lighting experiences in outdoor urban applications. The importance of lighting quality and psychological perception is a recent strand of research (So and Leung 1998; Boyce 2004). The psychological mechanism of lighting is complex and has been overlooked in research studies (Kim 2018; Veitch and Newsham 1996). It focuses on the visual indirect effects of lighting parameters (physical stimuli) that occur through the perceptual pathway and are able to stimulate subjective reactions

D. Casciani, *The Human and Social Dimension of Urban Lightscapes*,
PoliMI SpringerBriefs, https://doi.org/10.1007/978-3-030-57165-8_3

which can be physiological (change in physical properties), psychological (change in subjective experience) and behavioural (change in physical actions).

Lighting has already been proven to add meaning to the space by sending visual messages. These can be affective (affect, emotion, mood or subjective feelings), cognitive (thoughts, attention, imagination, perception), associative (memory, judgement) and motivational (proximity, openness, communication, engagement) (Steidle and Werth 2014; de Kort and Veitch 2014; Light + Behavior Symposium—Illuminating Engineering Society 2014; de Kort 2019) (Fig. 3.1).

Lighting allows the environmental affective evaluation (Flynn et al. 1973), the atmospheric evaluation (Vogels 2008) and also the environmental cognition and, as a contextual cue, results in cognitive associations which may influence behaviours (Casciani and Musante 2016). We selected applied and experimental lighting research studies that develop sets of interrelated constructs to specify relationships between variables, to explain and predict phenomena, and to build lighting theories (Veitch and Davis 2019). More than the results of how lighting variables can influence and correlate the perception of the environment in the applied and experimental lighting research (Chap. 2), in this chapter we discuss the methods available to conduct this kind of research.

To identify measures, we conducted a literature review examining established protocols, documents from international lighting bodies, and methods used in published and peer reviewed articles reporting subjective evaluation. In particular, the systematic review of the literature has focused on the methodologies, methods and tools used to assess how people responds psychologically (evaluation) or physically (behaviour) to the perceived lighted environment. In regards to the evaluation of the lit environment, we have included studies regarding the perceived quality of lighting,

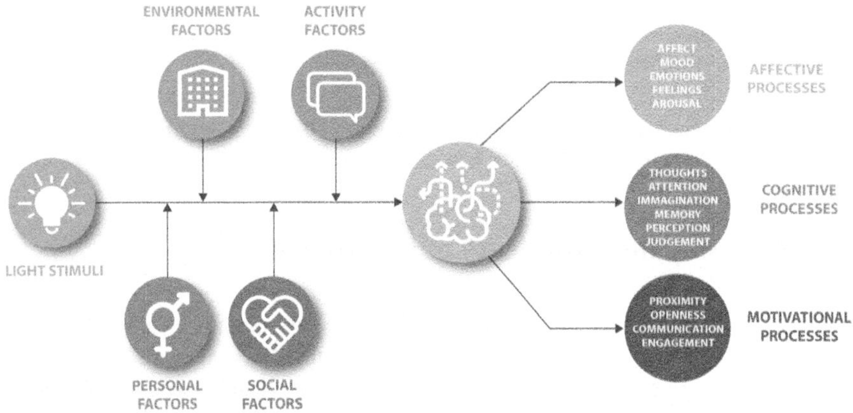

Fig. 3.1 Conceptual framework of the influence of several factors on social behaviours. The arrows represents the direction of the effect (Casciani 2020)

restorative experiences, the perceived safety and also the affective and social evaluation. Considering the influence on behaviours due to the lit environment, studies includes effects on walking, moving, seating, and socializing in the space.

The literature review has been mainly focused on understanding and comparing two main different kind of environmental assessment: the technical/objective and the subjective/observer-based one. The research procedures has been reviewed, highlighting the specific features related to data acquisition (direct elicitation and observation techniques) and data elaboration. In the latter part, the review has focused on how the lighting experiences were assessed in terms of in-situ real location, and laboratory installations, image and visualization based experiences, and immersive experiences through augmented reality (AR) technologies.

3.1.1 Objective Assessment of the Luminous Environment: Research Methods and Tools

The objective technical environmental assessment is "place-centred" and involves methods, tools and metrics to measure and read the environmental quality of a specific place. The recommendation of practice and normative of artificial lighting for public outdoor pedestrian areas state it is important to consider human needs in terms of obstacle detection, visual orientation, facial recognition and good perception of brightness along with general comfort and pleasantness (Fotios 2013; IES HANDBOOK, 10th Edition 2011). However, there is no general agreement on how the above human needs should be either operationalised or assessed. The objective evaluation of lighting typically involves detailed photometric measures of the environment: along with lighting levels and uniformity, advanced photometric measures include the necessity to capture the lighting information of the full environment.

One possibility is to analyse the luminance distribution (magnitude of present luminance within the space of investigation and research) to judge in objective and measurable way important lighting attribute such as the lighting distribution and the overall lighting quality. This analysis can be operationalized, using the high dynamic range (HDR) photography. This technique is based on an absolute point calibration using a luminance meter and a relative scene calibration using a calibrated camera and a HDR software to define the camera response curve of RGB pixel values to be translated in brightness. In addition to the luminance and lighting distribution, this image allow to conduct research on the probability to perceive glare, by using Evalglare in HDRScope (Wienold 2016). The evaluation of glare in pedestrian zones at night could be also conducted using UGR measures as explained by Villa et al. (2017).

3.1.2 Subjective Assessment of the Luminous Environment: Research Methods and Tools

Photometric assessments do not capture completely the lighting quality of a space, specifically from the subjective perspective. For this reason, the subjective environmental assessment is considered a viable assessment methodology because it is observer-based or person centred. It is based on self-reporting tools (Craik and Feimer 1987; Gifford 2007) to evaluate perceptions, observations and impressions of the laypersons related, in general, to the environmental quality (Bonaiuto e al. 2006; Fornara et al. 2010). In this paragraph, we explored specifically the methods and tools used to assess people judgement of an environment in relation to the presented lighting conditions and characteristics. Along with the innovative photometric approaches stated in the previous paragraph, different research methods and tools and models are used to address pedestrians' needs in relation to street lighting. These methods derive from both environmental psychology and social sciences and are adapted to the lighting research domain.

3.1.2.1 Emotional and Affective Appraisal Models

The Positive Affect Negative Affect Scale (PANAS) model is an emotion dimensional approach used to assess subjective appraisal (Watson et al. 1988). The positive dimension corresponds to emotion such as excited, inspired, proud and the negative scale is determined by emotions such as hostile, scared, guilty and nervous.

The Circumflex Model of Mehrabian and Russell (1974) uses the Pleasure-Arousal-Dominance (PAD) emotion model. It is used to recognize the emotional response of individuals and to predict the emotional influence on behaviours. This model distributes the emotions in two-dimensional spaces, defined as valence and arousal. Valence comprises pleasant and unpleasant emotions meanwhile arousal consists of a continuous state of neurological alertness and stimulation at different degrees, from no activation to extreme excitement. Arousal related to lighting is difficult to be demonstrated empirically and scientifically because it is a complex phenomenon which accounts for environmental and situational conditions other than solely lighting (Veitch and Newsham 1998). Due to different interpretations and limited measurements, arousal related to lighting processes are still poorly understood (Veitch 2001). Finally the dimension of Dominance consists in the amount of control over the circumstances.

3.1.2.2 Environmental Appraisal Models

Another approach to assess the lighting setting of a space (indoor or outdoor) is to evaluate the experience by means of perceived lighting impressions. Many studies (e.g. Veitch 2001; Loe et al. 1994) considered the research of Flynn (1977), and Flynn and Spencer (1977) the first to investigate the subjective appraisal and response of six lighting configurations in interior settings through the use of descriptive scales. The visual impact of lighting in the impression of a space has been investigated establishing an evident correlation between the appraisal of space and the design of certain lighting variables. In this studies, Flynn used 34 semantic differential scales through descriptive prompts such as "clarity", "order", "spaciousness", "relaxation" and "privacy". These scales allow a more impressionistic description of the lighting in a space and a characterization of the users affective response to the environment. The results were reduced to three main factors (perceptual clarity, evaluative impressions and spaciousness) through factor analysis. Furthermore, they applied multidimensional scaling to the ratings and elaborated an appraisal model for environmental lighting, consisting of uniform/non-uniform distribution of light, bright/dim levels of illumination, and overhead/peripheral lighting. The relation between these lighting dimensions and the impression dimension was then studied. Flynn's spatial-lighting design criteria resulted as an integration of some criteria proposed by Lynch (1960), Nasar (1998), Kaplan and Kaplan (1989) and are in accordance with the theories derived by lighting design practice of Richard Kelly (1952).

Another environmental appraisal model and metric is used by Vogel (2008) to measure the atmospheres. Differently from emotion or mood that assess an affective state, the atmosphere is the subjective evaluation of an experience of the lighted environments (Vogel 2008). The difference is that the atmosphere can have or cannot have an influence on one's affective state. Vice-versa, the perceived atmosphere of the environment is expected to be independent of people's emotions. For this reason, the perceived atmosphere is expected to be independent from people's mood and so more stable to determine the psychological effect of lit environments. A specific measurement tool was developed in form of a questionnaire of 38 descriptors aimed at quantifying the perceived atmosphere of lighted environments. Using factor analysis, Vogels found that the perceived atmosphere in an environment could be described by four factors: cosiness and liveliness, tenseness and detachment. This method and tool has been used to describe the luminous atmosphere. The relation between lighting parameters and perceived atmosphere has been also investigated, finding correlation between certain lighting features (CCT and intensity) and atmosphere perception.

Few studies have used subjective assessment research methods to explore the influence of lighting on the human and social experience of laypersons in built outdoor environments. Experimental studies already performed has not tried to capture more complex correlation and environmental qualities.

3.1.2.3 Direct Elicitation Techniques

The majority of the reviewed studies have been using direct elicitation techniques based on direct involvement of participants through surveys, interviews and verbal reports and conversations (e.g. Fiori et al. 2004). The questionnaires are based on a variety of semantic differentials (Likert rating scales) or pairwise comparisons to gather statistical information (descriptive and inferential) about the perception of the luminous atmosphere, investigating emotions, lighting impressions and the perceived atmosphere. These methodologies can be used to develop hypotheses on the correlated relationship (correlation method) between measured stimuli (lighting conditions) and observer's subjective responses (emotions or impressions). Research revealed different effect sizes with the paired-comparison design being more sensitive to detect the smallest differences with small groups of participants (e.g. Stokkermans et al. 2017b). Most studies used a one-dimensional method with observers correlating one single variable to a response, meanwhile others used multidimensional method to provide a series of different subjective judgements related to a lighting scenario.

Surveys could be designed as paper and pencil test or as a web-based tool. The first are vulnerable to wet conditions and need data transfer manually. In on-site outdoor assessment, web-based surveys could collect data less obtrusively using new technologies such as smart phones that could record assessed location and enable the researcher in doing some basic tabulation (Marans and Stokols 1993).

The analysis of data through rigorous statistical procedures and the statistical reporting toward accurate conclusions is an essential element of more reliable and scientific understanding of the results of the lighting research that use behavioural research methods. It is required to use parametric statistical tests, carried with a prior preliminary power analysis in order to justify sampling decisions. Besides, the effect size measurements are important to understand the effect magnitude and to encourage further studies and cumulative scientific knowledge (Uttley 2019).

3.1.2.4 Observation

With the on-site observation methods, the researcher observes people performing a task inside the natural environment. This passive observation prevents inadvertently interferences in the process, but does not provide any insight into the reason why certain behaviours are happening or certain decisions are made. Natural behaviours occurring in natural field of investigation can be used to assess either the current pattern of use due to the current lighted environment and to compare the differences after a lighting re-design. The majority of observations of the interpersonal communication in the various studies (e.g. Kobayashi 2013; Evensen 2014) consist of structured measurements accessing social proximity and social contacts both of verbal and non-verbal communications: latency (time between a question and an answer), talking time (percentage and length of conversation), talking sound level (decibel measurement), content of the conversation (general, abstract, intimate), eye contact (frequency and length), gaze direction, body direction and distance, interpersonal proximity (posture of the shoulder backwards or forward), gestures and facial expressions. In other studies, walking speed and path deviation of the participants' were observed to assess the influence of lighting on reassurance and safety perception in passing through residential paths (Muramatsu et al. 2001).

3.1.3 In Situ, in Scale and Virtual Lighting Experiences

For many years, the major topic in lighting research has been functionality with a focus on visual performance and visual comfort to carry out visual tasks efficiently, accurately and safely (Boyce 2003; Rea and Ouellette 1991; Veitch and Newsham 1998). The studies related to the assessment of affective, social appraisal and behaviours of lighting experiences has been mainly performed in interior spaces. Really few studies have explored the effects of lighting on the affective evaluation and behaviours in the urban environment. Reviewing the literature, we found that studies about the exploration of human appraisal and behaviours used prevalently in situ installations, both indoor (e.g. Miwa and Hanyu 2006; Taylor and Sucov 1974) and outdoor (e.g. Wänström Lindh 2012; Johansson et al. 2014). In these examples, it was possible to control and change the lighting variables along with furniture and decoration. Lighting in combination with the elements of the context, such as space dimensions and the features of the furniture, might influence particular meaning (e.g. Gifford 1988). Real-world studies are challenging due to the complexity of environments that are constituted of many other variables such as temperature, noise/sounds and people, making the research environment not easily controllable.

To increase the control over those many variables that could affect the reliability of the results, some studies were run in laboratory settings, through the redesign of realistic (e.g. Evensen 2014) or simplified spaces (e.g. Adams and Zuckermann 1991; Gergen et al. 1973; Rahm and Johansson 2018). If in these situation the control of different variables could be slightly simpler and the experience could be controlled better among participants, the overall test location might lack realism and so create bias in the results, that could not be generalized in realistic situation. The responses to lighting conditions could depend on the contextual situation and activities. Another hybrid possibility is to use Living Labs to conduct human centric research with direct involvement of stakeholders in real-life test beds. Living Labs are real environments studied and equipped to conduct a continuous series of experiments across different projects to generate insights about the influence of the use of light on human beings (Den Ouden et al. 2019).

Besides, some studies (e.g. Kobayashi et al. 2001) have used scale models for the evaluation of the appraisal of interiors and exteriors in order to compare different luminous situations through the change of different lighting parameters. This practice is generally more economic and easier to set up in terms of resources: time, money and recruited participants. It is mainly used in the preliminary stages of a research, to assess an idea of the correlation between evaluation and lighting features without setting up a full scale experiment. As a counter back, this method lacks in the realistic and immersive experience provided to the participants by the real 1:1 scale environments. Despite this, scale models are considered appropriate for simulating full-scale illuminated spaces and comparing different luminous conditions (Lau 1969, pp. 43–44). The results might predict a tendency that could be replicated in real, in situ or in laboratory, full scale experimental settings.

Another method is to assess the psychological mechanism of lighting through static photographic or rendered images and videos, to represent the different static or dynamic lighting conditions. Image-based experience are commonly used in environmental psychology for visual quality research. This method is always a trade-off between what is practical, what allows experimental control and what will generalize to the real situation (e.g. Nasar 1999, 2006; Stamps 2005). Images need to represent the world with the accuracy and the quality of the real one for more reliable results. In this regards, HDR (High Dynamic Range) photography are preferred in capturing the full dynamic of lighting contrast and brightness. To represent HDR images on standard dynamic range (SDR) displays, tone-mapped HDR images enable to render an accurate representation of perceived brightness within scenes using special compression algorithms (Inanici 2006). Computationally generated images are used to assess the perception and evaluation of either indoor and outdoor space's luminous atmosphere. This technique has been improved in the recent years through the use of rendering engines that allow the correct use and control of realistic lighting photometry toward the realization of physically correct and realistic images. Research studies based on rendered images of lighted spaces could show a very minimal representation of the interiors to avoid distractions (e.g. Stokkermans et al. 2017a); In other cases (e.g. Nasar and Bokharaei 2017a, b), extreme detailed images, including materials and furniture, increase the realism of the context to analyse also the contribution of elements other than solely lighting in the atmosphere perception.

The use of images has pros and cons. They are considered more direct and explanatory for picking up information on perception and impressions in comparison to descriptions; they represent a more immediate way to express visually what is impossible to explain with words (Gibson 1971). Gibson (1971) stated that

> "a picture that reconstitutes or represent the mosaic of colour sensations from an external scene will arouse the same process of perception that the external scene wold". On the other hand, "perception at second-hand will never be perception at first hand". (Gibson 1971)

If image based experiences might be lacking realism and not correlating appropriately with the results achievable in real and immersive conditions, previous research (e.g. Ulrich et al. 1991; Stamps 1990) has shown that, using coloured photographs and visualization, participant responses to such stimuli accurately reflect on-site responses. This is true also in the lighting research field where the same methodology was applied successfully to evaluate some real luminous scenes in comparison to pictures: many studies (e.g. Hendrick et al. 1977; Newsham et al. 2005; Villa and Labayrade 2012) have already stated that images are a reliable and valuable research tool, because there is a high correlation between preferences for images and preferences on-site. The comparison between real environment and visualized virtual environment indicated the validity and the effectiveness of such tools for lighting design, stating also that image based experience are more useful in testing lighting situation because they provide the possibility of manipulating lighting variables (e.g. Mahdavi and Eissa 2002; Eissa et al. 2001; Murdoch et al. 2015).

To evaluate the psychological and perceptive message of the lighting design installations, more appropriate methods were recommended to be used (Boyce 2004). In particular, behaviours influenced by lighting are advised to be assessed also with experimental studies through observations about the real behaviours of people under specific lighting conditions. Since the studies of behaviours regard how lighting can influence people actions, both laboratory and field research studies are suggested. In particular, field studies with overt behaviours are more effective because the experimental context is real (Boyce 2004); on the other hand, they can present uncontrollable variables that can make those studies unreliable and make the results contradictory.

New technology and innovative approaches can guarantee control over the experimental conditions, realism and immersion in research related to behaviours and lighting. This is clearly stated by Chamilothori et al. (2018) that conducted lighting research using immersive virtual reality technologies.

3.2 Conclusions

The research methods, procedures and tools used to capture human evaluation and behaviour in relation to indoor and outdoor lighting, provide a valuable perspective on how to deal with lighting research that include the perception of layperson and different stakeholders in considering, analysing or assessing different pedestrian lighting applications. Preference and perception depends on personal experiences and so it is extremely important to understand which are the operative modalities to capture multiple perspectives in assessing the lighted urban environment after dark. The understanding of the physical nocturnal environment by layperson is very important since is more holistic and could contrast or confirm the perception of architects and lighting designers (Bonnes et al. 2007).

Both the technical-objective and the subjective environmental assessments are aimed to achieve objectivity toward scientifically valid and reliable results (Craik and Feimer 1987). At the same time, they could both be considered subjective since also the selection of experimental locations and factors are based on human decisions. Furthermore, the two assessment approaches should be considered as complementary, used together with a multimethod approach to provide valuable cross-validation of the findings. This is particularly true when behaviours should be assessed: there is a need of combining these different techniques, both objective and subjective to merge the results and to evidence a correlation between what is stated by participants and what is effectively performed. Some studies (e.g. Evensen 2014) have shown that seldom, the results achieved by surveys were significant but in contrast with observed behaviours. In addition to this, it is important to perform physical measurements of lighting to have reliable and clear luminous conditions to be compared or replicated (Tiller 1990): luminance distribution of a wide areas, spectral power distribution perceived by eyes in adjunct to solely point illuminance (horizontal grid on surfaces or eye illuminance) and CCTs of the lighting fixtures.

In selecting which research methods and procedures are best fitting the scope of the research of correlating lighting variables to the influence on affective, social appraisal and behaviours, a balancing compromise needs to be done in order to achieve internal validity and also external validity. Controlled experimental situations to reduce noise might be counterbalanced by realism to not risk of producing environmental situation that are unlikely to exist out of the laboratory settings (Veitch and Davis 2019).

Besides the stated relevance for research, these procedures for assessing the nocturnal environment in terms of affective, social appraisal and behaviours of human scale lighting experience has also many implications for practice. Research based on this procedure can help the designers and practitioners in understanding how they can combine different lighting variables, and luminaire arrangements to create a certain atmosphere in a space. A good designed luminous solutions needs to accomplish either the functional/technical scopes such as energy efficiency, reliability and safety along with the emotional scopes that deal with the human and social response to the luminous installation. The systematic use of objective and subjective assessment of outdoor lighting before, during and after a project of lighting design could help educate lighting designers and decision-makers toward more human-oriented lighting practice.

References

Adams L, Zuckerman D (1991) The effect of lighting conditions on personal space requirements. J Gen Psychol 118(4):335–340. https://doi.org/10.1080/00221309.1991.9917794

Bonaiuto M, Fornara F, Bonnes M (2006) Perceived residential environment quality in middle- and low-extension Italian cities. Eur Rev Appl Psychol 56(1):23–34. https://doi.org/10.1016/j.erap.2005.02.011

Bonnes M, Uzzell D, Carrus G, Kelay T (2007) Inhabitants' and experts' assessments of environmental quality for urban sustainability. J Soc Issues 63(1):59–78. https://doi.org/10.1111/j.1540-4560.2007.00496.x

Boyce PR (2003) Human factors in lighting. Taylor & Francis, London

Boyce PR (2004) Lighting research for interiors: the beginning of the end or the end of the beginning. Lighting Res Technol 36(4):283–293. https://doi.org/10.1191/1147715350 4li118oa

Casciani D (2020) The psycho-social influence of lighting in space perception: an experimental proof of concept. Int J Archit Spat Environ Design 14(2):15–37. https://doi.org/10.18848/2325-1662/cgp/v14i02/15-37

Casciani D, Musante F (2016) What light does: reflecting on the active social effects of lighting design and technology. In: Sociotechnical environments: proceedings of the 6th STS Italia conference, STS Italia Publishing, Trento

Chamilothori K, Wienold J, Andersen M (2018) Adequacy of immersive virtual reality for the perception of daylit spaces: comparison of real and virtual environments. LEUKOS 15(2–3):203–226. https://doi.org/10.1080/15502724.2017.1404918

Craik K, Feimer N (1987) Environmental assessment. In: Stokols D, Altman I (eds) Handbook of environmental psychology. Wiley, New York, pp 891–918

de Kort Y (2019) Tutorial: theoretical considerations when planning research on human factors in lighting. LEUKOS 15(2–3):85–96. https://doi.org/10.1080/15502724.2018.1558065

de Kort Y, Veitch J (2014) From blind spot into the spotlight. J Environ Psychol 39:1–4. https://doi.org/10.1016/j.jenvp.2014.06.005

Den Ouden E, Haans A, Ross P, Van Essen H, De Kort Y, Brankaert R (2019) 10 years living light labs 2019 achievements and learnings of researching intelligent lighting solutions in living labs at the TU/e Intelligent Lighting Institute. ILI Intelligent Lighting Institute, TU/e, Eindhoven

Eissa H, Mahdavi A, Klatzky R, Siegel J (2001) Computationally rendered architectural spaces as means of lighting design evaluation. In: de Vries B, van Leeuwen J, Achten H (eds) Computer aided architectural design futures 2001. Springer, Dordrecht. https://doi.org/10.1007/978-94-010-0868-6_26

Evensen I (2014) A study on the effects of lighting on social interaction (undergraduate). The Bartlett School of Graduate Studies

Fiori S, Houdemont K, Leroux M (2004) Parcours nocturnes: étude sociologique sur l'ambiance et les usagers nocturnes Mail d'Empalot et cité Bordelongue (Toulouse). Rapport de recherche 61 hal-01364547f, Cresson, p 64

Flynn J (1977) A study of subjective responses to low energy and nonuniform lighting systems. Lighting Design Appl 7:6–15

Flynn J, Spencer T (1977) The effects of light source color on user impression and satisfaction. J Illum Eng Soc 6(3):167–179. https://doi.org/10.1080/00994480.1977.10747811

Flynn J, Spencer T, Martyniuk O, Hendrick C (1973) Interim study of procedures for investigating the effect of light on impression and behavior. J Illum Eng Soc 3(1):87–94. https://doi.org/10.1080/00994480.1973.10732231

Fornara F, Bonaiuto M, Bonnes M (2010) Cross-validation of abbreviated perceived residential environment quality (PREQ) and neighborhood attachment (NA) indicators. Environ Behav 42:171–196. https://doi.org/10.1177/0013916508330998

Fotios S (2013) LRT digest 1. Maintaining brightness while saving energy in residential roads. Lighting Res Technol 45:7–21. https://doi.org/10.1177/1477153512464141

Gergen J, Gergen M, Barton H (1973) Deviance in the dark. Psychol Today 7(5):129–131

Gibson J (1971) The information available in pictures. Leonardo 4(1):27. https://doi.org/10.2307/1572228

Gifford R (1988) Light, decor, arousal, comfort and communication. J Environ Psychol 8(3):177–189. https://doi.org/10.1016/s0272-4944(88)80008-2

Gifford R (2007) Environmental psychology: principles and practice, 4th edn. Colville, Colville

Hendrick C, Martyniuk O, Spencer T, Flynn J (1977) Procedures for investigating the effect of light on impression: simulation of a real space by slides. Environ Behav 9(4):491–510. https://doi.org/10.1177/001391657794003

IES handbook, 10th edn (2011)

Inanici M (2006) Evaluation of high dynamic range photography as a luminance data acquisition system. Lighting Res Technol 38(2):123–134. https://doi.org/10.1191/1365782806li164oa

Johansson M, Pedersen E, Maleetipwan-Mattsson P, Kuhn L, Laike T (2014) Perceived outdoor lighting quality (POLQ): a lighting assessment tool. J Environ Psychol 39:14–21. https://doi.org/10.1016/j.jenvp.2013.12.002

Kaplan R, Kaplan S (1989) The experience of nature: a psychological perspective. Cambridge University Press, Cambridge

Kelly R (1952) Lighting as an integral part of architecture. College Art J 12(1):24. https://doi.org/10.2307/773361

Kim D (2018) Development of a psychological pathway model linking lighting quality to well-being in indoor café environments. Indoor Built Environ 27(3):390–401. https://doi.org/10.1177/1420326x17753512

Kobayashi S (2013) Lighting effects on unconscious human behaviors. In: 10th International symposium on sustainable healthy buildings annual spring conference. Society of Lighting & Visual Environment, Seoul

Kobayashi S, Inui M, Nakamura Y (2001) Preferred illuminance non-uniformity of interior ambient lighting. J Light Visual Environ 25(2):64–75. https://doi.org/10.2150/jlve.25.2_64

Lau J (1969) Differences between full-size and scale-model rooms in the assessment of lighting quality. In: Architectural psychology. Proceedings of the conference, Dalanhui, University of Strathclyde

Light + Behavior Symposium—Illuminating Engineering Society (2014) Retrieved 17 May 2020, from https://www.ies.org/research/events/research-symposia/light-behavior-symposium/

Loe L, Mansfield K, Rowlands E (1994) Appearance of lit environment and its relevance in lighting design: experimental study. Lighting Res Technol 26(3):119–133. https://doi.org/10.1177/096032719402600301

Lynch K (1960) The image of the city. The MIT Press, Cambridge

Mahdavi A, Eissa H (2002) Subjective evaluation of architectural lighting via computationally rendered images. J Illum Eng Soc 31(2):11–20. https://doi.org/10.1080/00994480.2002.10748388

Marans RW, Stokols D (1993) Environmental simulation: research and policy issues. Plenum, New York

Mehrabian A, Russell J (1974) The basic emotional impact of environments. Percept Motor Skills 38(1):283–301. https://doi.org/10.2466/pms.1974.38.1.283

Miwa Y, Hanyu K (2006) The effects of interior design on communication and impressions of a counselor in a counseling room. Environ Behav 38(4):484–502. https://doi.org/10.1177/0013916505280084

Muramatsu R, Nakamura Y, Nakajima S, Kobayashi S (2001) Effects of house exterior lighting on the evaluation of lighting environment on nighttime residential streets. J Light Vis Environ 25(1):60–65. https://doi.org/10.2150/jlve.25.1_60

Murdoch M, Stokkermans M, Lambooij M (2015) Towards perceptual accuracy in 3D visualizations of illuminated indoor environments. J Solid State Lighting 2(1). https://doi.org/10.1186/s40539-015-0029-6

Nasar JL (1998) The evaluative image of the city. Sage, Thousand Oaks

Nasar JL (1999) Design by competition: making design competition work. Cambridge University Press, New York

Nasar J (2006) Design by competition: making design competition work. Cambridge University Press, Cambridge

Nasar JL, Bokharaei S (2017a) Lighting modes and their effects on impressions of public squares. J Environ Psychol 49:96–105

Nasar JL, Bokharaei S (2017b) Impressions of lighting in public squares after dark. Environ Behav 49(3):227–254

Newsham GR, Richardson C, Blanchet C, Veitch JA (2005) Lighting quality research using rendered images of offices. Lighting Res Technol 37(2):93–112. https://doi.org/10.1191/1365782805li132oa

Rea MS, Ouellette MJ (1991) Relative visual performance: A basis for application. Light Res & Technol 23(3): 135–144. https://doi.org/10.1177/096032719102300301

Rahm J, Johansson M (2018) Assessing the pedestrian response to urban outdoor lighting: a full-scale laboratory study. PLoS ONE 13(10):e0204638. https://doi.org/10.1371/journal.pone.0204638

So A, Leung L (1998) Indoor lighting design incorporating human psychology. Archit Sci Rev 41(3):113–124. https://doi.org/10.1080/00038628.1998.9697420

Stamps AE (1990) Use of photographs to simulate environments. A meta-analysis. Percept Motor Skills 71:907–913

Stamps AE (2005) Enclosure and safety in urbanscapes. Environ Behav 37(1):102–133. ISSN 0013-9165

Steidle A, Werth L (2014) In the spotlight: brightness increases self-awareness and reflective self-regulation. J Environ Psychol 39:40–50. https://doi.org/10.1016/j.jenvp.2013.12.007

Stokkermans M, Vogels I, de Kort Y, Heynderickx I (2017a) Relation between the perceived atmosphere of a lit environment and perceptual attributes of light. Lighting Res Technol 0:1–15. https://doi.org/10.1177/1477153517722384

Stokkermans M, Vogels I, de Kort Y, Heynderickx I (2017b) A comparison of methodologies to investigate the influence of light on the atmosphere of a space. LEUKOS. https://doi.org/10.1080/15502724.2017.1385399

Taylor LH, Sucov EW (1974) The movement of people toward lights. J Illum Eng Soc 3:237–241

Tiller DK (1990) Toward a deeper understanding of psychological aspects of lighting. J Illum Eng Soc 19(2):59–65. https://doi.org/10.1080/00994480.1990.10747963

Ulrich RS, Simons RF, Losito BD, Fiorito E, Milest MA, Zelson M (1991) Stress recovery during exposure to natural and urban environment. Academic Press Limited

Uttley J (2019) Power analysis, sample size, and assessment of statistical assumptions—improving the evidential value of lighting research. LEUKOS 15(2–3):143–162. https://doi.org/10.1080/15502724.2018.1533851

Veitch JA (2001) Psychological processes influencing lighting quality. J Illum Eng Soc 30(1):124–140

Veitch JA, Newsham G (1996) Determinants of lighting quality II: research and recommendations. In: 104th Annual Convention of the American Psychological Association. Toronto, Ontario, Canada pp 1–38

Veitch JA, Davis RG (2019) Lighting research today: the more things change, the more they stay the same. LEUKOS 15(2–3):77–83. https://doi.org/10.1080/02604027.2019.1596680

Veitch JA, Newsham GR (1998) Determinants of lighting quality I: state of the science. J Illum Eng Soc 27(1):92–106. https://doi.org/10.1080/00994480.1998.10748215

Villa C, Labayrade R (2012) Validation of an online protocol for assessing the luminous environment. Lighting Res Technol 45:401. Originally published online 22 June 2012

Villa C, Bremond R, Saint-Jacques E (2017) Assessment of pedestrian discomfort glare from urban LED lighting. Lighting Res Technol [Online] 49(2):147–172. https://doi.org/10.1177/1477153516673402

Vogel I (2008) Atmosphere metrics: a tool to quantify perceived atmosphere. In: Probing experience. From assessment of user emotions and behaviour to development of products, 1st edn. Springer, The Netherlands

Wänström Lindh U (2012) Light shapes spaces experience of distribution of light and visual spatial boundaries. Thesis for the degree of Doctor of Philosophy in Design at HDK, School of Design and Crafts, Faculty of Fine, Applied and Performing Arts, University of Gothenburg. ISBN: 978-91-979993-2-8

Watson D, Clark LA, Tellegen A (1988) Development and validation of brief measures of positive and negative affect: the PANAS scales. J Pers Soc Psychol 54(6):1063–1070

Wienold J (2016) Evalglare—determines and evaluates glare sources [Online], 15 June 2016. Radiance Online. Available from: https://www.radiance-online.org//learning/documentation/manual-pages/pdfs/evalglare.pdf/view

Chapter 4
In-Situ Experience of the Human-Scale Urban Lightscape

Abstract This chapter presents a procedure for the analysis, design and post-installation evaluation of the reciprocal relationship between lighting, city and people. Through a qualitative and quantitative subjective assessment, it also reports the results of the influence of lighting variables such as the correlated colour temperature (CCT) and lighting spatial distribution in terms of luminous environmental atmosphere (comfort, liveliness, safety, interest, pleasure), social appraisal (sociability), lighting appearance (brightness, uniformity, warm/cold CCT) and spatial appraisal (dimensions, coherency, complexity, mystery). The research provides an evidence-based correlation of lighting variables to more human- and social-oriented atmospheres.

Keywords Urban lighting design · Lighting appearance · Luminous environmental atmosphere · LEDs · Lighting design · Social appraisal · Spatial appraisal

4.1 Introduction

This research is based on the hypothesis that lighting could influence social appraisal and social behaviours in the outdoors because it allows for environmental cognition at night-time, influences the impression and the affective evaluation of the atmosphere of a space, and, as a contextual cue, defines cognitive associations, which may motivate people. Therefore, the aim of this work was to study and compare different lighting conditions in a real, pedestrian, semi-outdoor environment, by evaluating people's opinions and responses. It was expected that certain lighting variables could influence and define a more human- and social-oriented urban environment. The evaluation of urban-luminous experiences from people's perspectives (see Chap. 3) has been hypothesized to be crucial for designing night-time environments for the benefit of the social use of the public realm. With this objective, a procedure for analysing, designing and evaluating the reciprocal relationship between lighting and people was tested in a real "*socio-physical unit of environmental experience*" (Canter 1977). The research questions addressed in this study were the following:

D. Casciani, *The Human and Social Dimension of Urban Lightscapes*,
PoliMI SpringerBriefs, https://doi.org/10.1007/978-3-030-57165-8_4

- How can lighting design encourage the positive use of the space?
- How can lighting design contribute to making an urban environment a safer, more pleasant and more attractive space?
- How can lighting design evoke interest in space?
- Which lighting features can positively impact the perceived atmosphere of the night space?

In this regard, the research was focused on understanding people's appraisal of different factors (luminous atmosphere, lighting appearance, sociality and spatiality) in relation to different lighting variables: for instance, warm and cold CCT, lighting distribution between uniform and non-uniform lighting, and the juxtaposition of brightness and darkness. From this, more specific research questions arose and focused on

- What is the effect of perceived brightness and darkness in the experience of the atmosphere of the space?
- What is the effect of CCT on the perceived atmosphere of the space?
- What is the effect of spatial lighting distribution (namely, uniformity vs. not uniformity) on the perceived atmosphere of the space?

From the preliminary literature review and also taking into account the insights gathered from the practice of lighting design, some general hypotheses were proposed. Starting from the general assumption that the differences between lighting scenarios were expected to be recognized, uniform lighting was expected to be correlated with safety perception; meanwhile, non-uniform lighting was expected to be perceived as less safe but more intriguing and mysterious due to the brightness contrast. In this, orderly lighting scenarios with lighting rhythmically enhancing the architecture were expected to be more stimulating and interesting in comparison to irregular, non-uniform and uniform lighting. The use of warm lighting was expected to contribute to the perception of comfort and sociability.

4.2 Methodology

4.2.1 Description of the Space and the Lighting System

The experiments took place from November 2012 to March 2013 at the Market Hall Living Lab (den Ouden et al. 2019), a space conceived as an environmental laboratory for testing intelligent lighting solutions and human-centric perspectives. Located in front of the Metaforum at the Eindhoven University of Technology (TU/e) campus, the space (62.5 m × 75 m) is a roofed and not completely enclosed area, which works both as a connective hub for staff and student transitions, and also as a public meeting area, housing many activities and events. Redesigned by Ector Hoogstad Architecten, it represents a historical icon of the university, a flexible and hybrid space standing at the threshold between indoor and outdoor spaces (Fig. 4.1).

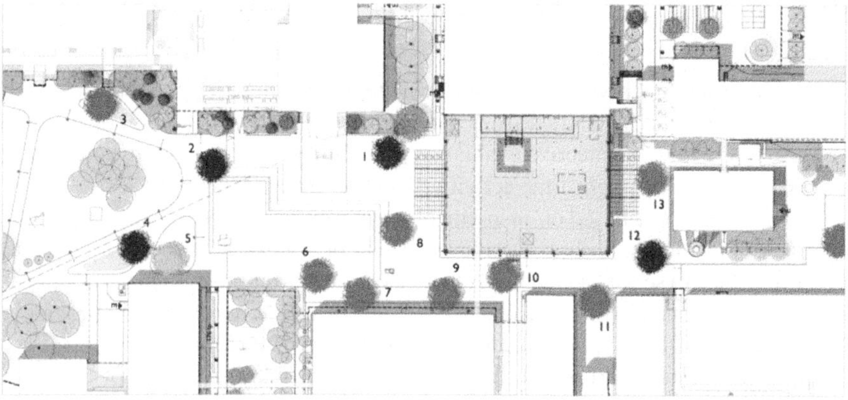

Fig. 4.1 Plan view (in grey) of the Market Hall Living Lab of the Tu/e. *Credits* Ector Hoogstad Architecten (2012)

The permanent lighting infrastructure of the building is composed of warm and cool ceiling luminaires, with LED modules, in a 20 by 24 grid. The 480 LED modular fixtures are located at about 10 and 12 m in the highest zones and at about 3 m in the lowest zones in correspondence to the main entrance. The LED modular system (LMS) is equipped with 4 power LEDs and a TIR lens, with narrow (8°) and medium (36°) beams. Each LMS is individually controllable in terms of spatial distribution, correlated colour temperature (CCT) of white lighting between cold (6000 K) and warm (3000 K) CCT, and lighting intensities, with a DMX controller (Fig. 4.2).

Fig. 4.2 Lighting infrastructure and variable spatial lighting distribution

4.2.2 Participants and Procedure

The research approach was based on a two-steps procedure: (i) the analysis of the space to understand the context from an objective and subjective point of view and (ii) the objective and subjective evaluation of lighting design alternatives that were designed considering night-time activities and the needs of the users.

The first phase was based on an in-situ observation of the "as-it-is" situation during a period of one month, in short sessions of one hour per night. This activity included measuring the presence of people in the space, observing movements, monitoring the timing of the use of the space and assessing recurrent behaviours to provide an understanding of how people made use of the space. A preliminary interview of about 40 people was performed to collect impressions and meanings of the space, garnering their general opinions about the Market Hall by asking, "*Are you familiar with this place? Do you like this space?*".

The second experimental phase was performed to evaluate the atmospheric impressions and the lighting influence on people through a qualitative and quantitative approach. Borrowing techniques from environmental psychology to assess the subjective impression of luminous visual environments, a quantitative data collection was performed, followed by a qualitative semi-structured interview. A questionnaire about the influence of lighting was administered to gather people's evaluations of four factors: luminous environmental atmosphere (comfort, liveliness, security and safety, interest, pleasure), social appraisal (sociability), lighting appearance (brightness, uniformity, warm/cold CCT) and spatial appraisal (dimensions, coherency, complexity, mystery).

The luminous environmental atmosphere was based on the atmospheric questionnaire of Vogels' (2008), and it was implemented using spatial descriptors derived from Kaplan and Kaplan (1989). Voluntary participants in the experiment were recruited directly in the space of the Market Hall. They were invited to participate in a survey, followed by a semi-structured interview, to assess their impression of the Market Hall atmosphere. After a 10-min explanation of the study, which allowed the eyes to adapt to outdoor lighting levels, participants were asked to answer a questionnaire lasting approximately 10 min, which was based on observing alternative lighting scenarios in pairs. Through a pairwise comparison, participants answered the following questions: "*Under which lighting condition, the space appears more …?*" in relation to the items "*sociable*", "*cosy*", "*lively*", "*safe*", "*interesting*", "*pleasant*", "*bright*", "*uniform*", "*warm*", "*spacious*", "*coherent*", "*complex*" and "*mysterious*". To better understand the meaning of the items, participants were provided with supplementary written explanations in English and Dutch. At the end of the survey, participants were interviewed to explain and discuss the reasons for their choices. The sample of the participants in this phase was composed of 30 subjects divided into three groups, each consisting of 10 participants. The total sample was composed mainly of students (83.3%); the majority were men (73.3%) and only a small percentage were women (26.7%). The average age was 23.3 ± 3.9 years old. The sample was mainly composed of Dutch participants (60%), then Chinese (13.3%) and Turkish

(6.7%) participants, and the rest were Bhutanese, Costa Rican, Ethiopian, Greek, German and Romanian (20%). The majority of respondents (97%) did not have any knowledge about lighting, neither at a student nor a professional level.

4.2.3 Lighting Scenarios: Design and Description

Six alternative lighting scenarios were designed, taking into account the research questions and the lighting infrastructure available. The horizontal and vertical illuminance (E_v and E_h) values were set up to be compliant to the normative and codes of practice (DiLaura et al. 2011; BSI 5489.1:2003 + A2: 2008), related to the lighting of squares and semi-opened public/private urban spaces. The experiment was conducted, comparing in pairs six different lighting modes: A1 (carpet of light warm), A2 (carpet of light cold), B1 (peripheral background), B2 (central foreground), C1 (layered orderly), C2 (layered disorderly). Carpet of light (A1–A2) was designed to ensure uniform and overhead lighting in the whole space: the two compared lighting scenarios presented two opposite CCT (3000 K and 6000 K); meanwhile, they had the same average E_v and E_h. Uniform lighting refers to ambient lighting (Kelly 1952) and is supposed to enable the users to get their bearing, providing a feeling of safety. B1–B2 lighting conditions were designed to create non-uniform lighting distribution in different functional zones of the space, ensuring that visual contact points of entrance and exit (gates) were always lit. The lighting scenarios were enhancing the peripheral background (B1) or the central foreground (B2) and showed a contrast between lit and dark areas. Finally, the layered approach was defined by general dimmed warm ambient lighting, and high-intensity accents focused on different architectural and functional elements of the space such as the perimeter, the benches and the columns. The accent lighting was designed in an orderly way with a neutral CCT (C1) or in a disorderly way with mixed warm and cold CCTs (C2). The dark-and-light juxtaposition refers to focal glow (Kelly 1952), useful to highlight visual cues, to emphasize specific areas, to create hierarchies of perception, and to attract attention. In Table 4.1, the lighting conditions are represented through tonemapped images, luminance images, and measured values. The assessment of environmental luminance was conducted through High Dynamic Range (HDR) photographic images. A Canon Eos 550D camera equipped with a standard lens (18–55 mm) was used for the production of the images. These were subsequently processed by means of two open-source software programs: Luminance HDR (2019) for tone mapping and HDRscope ("HDRscope" 2019; Kumaragurubaran and Inanici 2013) for deriving luminance maps that enable the understanding of the lighting distribution in the environment (Decuypere et al. 2009; Inanici 2006).

Table 4.1 Description of the lighting conditions and measured features

Lighting scenario	Tonemapped and luminance images (from the point of view of the respondents)	Lighting features
Carpet of light		A1: uniform warm CCT (3000 K). E_h max is 15.8 lx - E_h min is 8 lx. E_v is average 4 lx
		A2: uniform cold CCT (6000 K). E_h max was 16 lx while E_h min was 8.5 lx. E_v is average 4 lx
Lighting zones		B1—Peripheral Background: neutral CCT in the main entrance and perimeters: $E_{h\ max}$ of 20 lx and $E_{h\ min}$ 10 lx with $E_{v\ max}$ 10 lx (entrance)
		B2—Central Foreground: neutral CCT and lit pathways with lighting levels the same as the previous scenario
Layered approach		C1—Orderly: general illuminance $E_{h\ max}$ 7 lx and a $E_{h\ min}$ 5 lx with $E_{h\ average}$ of 6 lx. High intensity accents have $E_{h\ max}$ of 80 lx

(continued)

Table 4.1 (continued)

Lighting scenario	Tonemapped and luminance images (from the point of view of the respondents)	Lighting features
		C2—Disorderly: same lighting levels of the C1 with different CCT for the accent lighting. Random cold and warm CCT on the columns and cold CCT for the vegetation (perimeters)

4.2.4 Analysis Methodology

The first part of the research was based on observations, complemented with the analysis of integration, visibility and connectivity with the software Depthmap X ("Software | Space Syntax Network" 2019) to understand the patterns of movements and visual cues (Hillier and Hanson 1984).

The second step of the procedure was evaluated in a quantitative (questionnaires) and qualitative (interviews) way. The experiment was conducted on a sample of 30 subjects, divided into three groups (10 per group), for the analysis of comparisons A1 versus A2 and A1 versus B1, B1 versus B2 and B1 versus C1, C1 versus C2 and C1 versus A1.

The collected data from the questionnaires were statistically analysed with the software STATA15 (Stata Corp., TX, USA). For the descriptive statistics of the sample, mean and standard deviation were used for quantitative data. Differently, percentages and frequencies were considered for qualitative insights, supported by the audio-recorded interviews, which were re-written, analysed and plotted in clusters. The homogeneity tests between the different groups were performed with the Pearson test goodness of fit; statistical significance was set at 5% ($p < 0.05$). The comparisons of the different scenarios were made by analysing the percentage values, without statistical inference tests, given the composition of the dataset.

The combination of quantitative statistical data and qualitative insights was aimed to derive a more complete and clear understanding of the relationship between lighting and people.

4.3 Data Analysis and Results

4.3.1 Pedestrian Paths and Night-Time Activities

The Visibility Graph Analysis (VGA) was run to analyse the level of visibility of the space and to investigate whether the morphology of the built environment itself influences pedestrian movement through space, by comparing social and cultural formations via spatially organizing constructs. About connectivity, the Market Hall is placed in a good position: it is the widest and fastest way to reach the different parts of the Campus through the many gates, which increases the probability pedestrians will pass through the zone (Figs. 4.3 and 4.4). The Space Syntax analysis considers a daily-based view of the space, with natural lighting conditions that define the perception of the surrounding. By night, the perception and navigation of the Market Hall could be different: in low lighting levels, the human eyes are driven to the most luminous areas. In addition to this, the observations confirmed the instrumental use of the space by people passing through the main gates and using the shortest trajectories. A night-time activities diagram was built (Fig. 4.5), considering the links among night-time hours, activities and pattern of use, assuming six social moments in the night of the Market Hall:

- POST-WORK: as the sun sets, depending on the season, either the workday extends into the night, or daylight extends into the post-work; mainly late employees use the space;
- HAPPY HOUR/DINING OUT: the social extension of the workday, with the decompression time of Happy Hour every Thursday, longer shopping timing on Friday, or eventually people getting food from outside and eating in the Library;

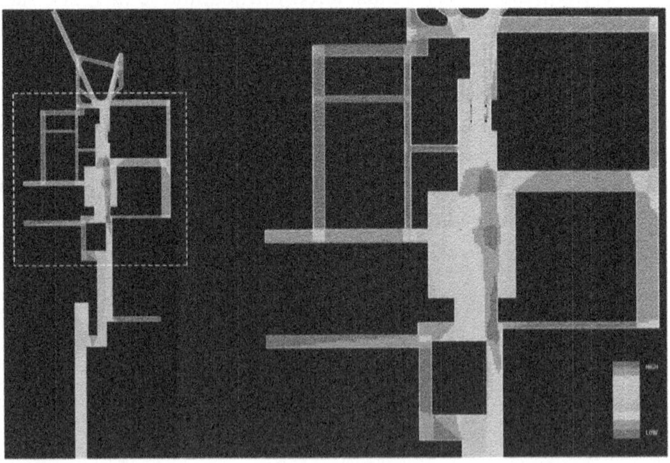

Fig. 4.3 Level of Connectivity of spaces indicating where people are moving given the spatial structure

- BACK HOME: the transit of "the late students" back to their homes;
- NIGHT SHIFT: the area is mostly empty;
- EARLY USERS: the first workers arrive in the space; and
- WORK TIME: students and university staff arrive back to school and work.

The activities occurring in this space during the night are used as a framework to design the lighting conditions accordingly.

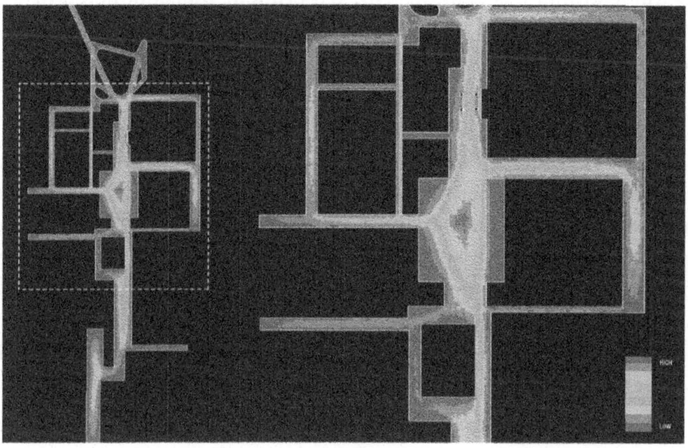

Fig. 4.4 Gate counts state that Market Hall is the core center of the passages from the surrounding areas

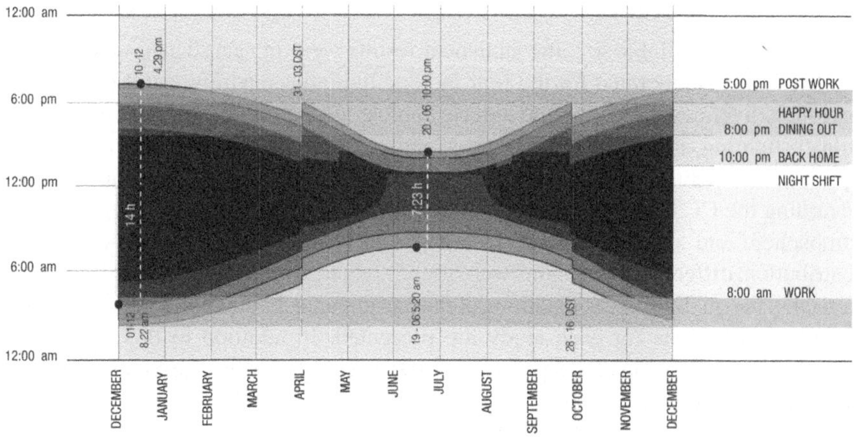

Fig. 4.5 Annual night-time activities map for the Market Hall

4.3.2 People's Expectations and Appraisal of the Space

The Market Hall was perceived as a blank canvas with a "*wide*", "*semi-empty*", "*anonymous*" and "*underexploited*" character, lacking both people presence and social activities. From the survey, it emerged as being used exclusively in a functional way: the perception of underuse derived both by low social occurrences and missing cultural or historical references. The low social use could derive also by the unfavourable climatic conditions (the experiment was conducted in winter, with low temperatures between −11 and 5 °C and adverse climatic conditions such as snow, rain and wind). The Market Hall was perceived as senseless and unfinished, "*sober, still, boring*" with materials and architectural elements that reminded participants of an "*industrial*", "*heavy structure*". It was considered a "*non-place*", lacking relational, historical and identity aspects (Augé 1992). The area was mainly used as an exchange zone for transition in the very early morning and late evening. The respondents were all very familiar with the space (95%); being usual commuters, they confirmed they used the space "*just as a passage*". The general appraisal for the outdoor area was positive (86.6%); on the other hand, 13.4% considered the space as not pleasant and appreciable, but potentially improvable with "*furniture and greenery*". In addition to this, 83% of participants seemed to be very interested in and curious about the possibilities offered by the lighting system.

4.3.3 Lighting Design Affecting the Perception and Sociality of the Space

As represented in Table 4.2, the statistical results were in accordance with most of the hypotheses elaborated for the four factors: luminous environmental atmosphere (comfort, liveliness, security and safety, interest, pleasure), social appraisal (sociability), lighting appearance (brightness, uniformity, warm/cold CCT) and spatial appraisal (dimensions, coherency, complexity, mystery). The main findings are regarding the CCT and its influence on impressions of the luminous environmental atmosphere and social appraisal. In relation to uniformity versus non-uniformity distribution, different impressions were found statistically significant but also controversial in the qualitative explanations of the participants. In the following paragraphs, the obtained results for each study are presented. In addition to this, qualitative descriptions supporting and explaining quantitative data are discussed, related to the following issues:

- The perception of sociability and the warm CCT of lighting
- Lighting distribution: functional versus evocative
- The spatial perception: brightness and uniformity perception
- The spatial perception: understanding the space through its visual cues.

Table 4.2 Statistical elaboration of data with dark grey $p < 0.01$ and clear grey $p < 0.05$ level of significance

Study	Comparison	Lighting Scenario	Luminous environmental atmosphere					Social appraisal	Lighting appearance			Spatial appraisal			
			COSY	SAFE	LIVELY	INTERESTING	PLEASANT	SOCIABLE	BRIGHT	UNIFORM	WARM	SPACIOUS	COHERENT	COMPLEX	MYSTERIOUS
1	1	A1	100	50	100	60	90	100	20	10	100	20	80	20	60
		A2	0	50	0	40	10	0	80	90	0	80	20	80	40
		p	0.000	1.000	0.000	0.526	0.007	0.000	0.050	0.007	0.000	0.050	0.050	0.050	0.526
	2	A1	90	50	70	30	70	80	40	70	100	50	80	20	40
		B1	10	50	30	70	30	20	60	30	0	50	20	80	60
		p	0.007	1.000	0.200	0.200	0.200	0.050	0.526	0.200	0.000	1.000	0.050	0.050	0.526
2	1	B1	40	50	50	50	40	50	40	60	40	60	50	50	60
		B2	60	50	50	50	60	50	60	40	60	40	50	50	40
		p	0.526	1.000	1.000	1.000	0.526	1.000	0.526	0.526	0.526	0.526	1.000	1.000	0.526
	2	B1	50	0	10	60	50	10	0	0	60	10	0	40	70
		C1	50	100	90	40	50	90	100	100	40	90	100	60	30
		p	1.000	0.000	0.007	0.526	1.000	0.007	0.000	0.000	0.526	0.007	0.000	0.526	0.200
3	1	C1	40	90	80	20	60	80	100	80	40	80	50	50	10
		C2	60	10	20	80	40	20	0	20	60	20	50	50	90
		p	0.526	0.007	0.050	0.050	0.526	0.050	0.000	0.050	0.526	0.050	1.000	1.000	0.007
	2	A1	90	60	70	80	100	80	30	80	100	60	70	50	70
		C1	10	40	30	20	0	20	70	20	0	40	30	50	30
		p	0.007	0.526	0.200	0.050	0.000	0.050	0.200	0.050	0.000	0.526	0.200	1.000	0.200

4.3.3.1 Study 1: Comparisons A1 Versus A2 and A1 Versus B1

Considering the first group of participants, multiple statistical significances emerged. In the comparison A1 versus A2, the lighting condition A1 was considered sociable, lively, cosy and warm by 100% of the participants ($p < 0.01$), pleasant by 90% ($p < 0.01$) and coherent by 80% ($p < 0.05$). The condition A2, instead, was defined as complex, spacious and bright by 80% ($p < 0.05$) and uniform by 90% ($p < 0.01$) of people. In the comparison A1 versus B1, the condition A1 was perceived as warm by 100% of participants ($p < 0.01$), cosy by 90% ($p < 0.01$), and sociable and coherent by 80% ($p < 0.05$) of respondents. The B1 condition was perceived as complex by 80% of subjects ($p < 0.05$). In terms of safety, none of the lighting conditions were considered safe or unsafe.

4.3.3.2 Study 2: Comparisons B1 Versus B2 and B1 Versus C1

Considering the second group, no statistical difference, but rather, homogeneity between perceptions regarding the two conditions, B1 versus B2, emerged. In the comparison B1 versus C1, numerous significances at the level of 1% emerged. In particular, the condition C1 was defined as coherent, safe, uniform and bright by 100% ($p < 0.01$) of the participants, and sociable, spacious and lively by 90% ($p < 0.01$) of the respondents.

4.3.3.3 Study 3: Comparisons C1 Versus C2 and A1 Versus C1

The third group of participants perceived the C1 condition as sociable (80%, $p < 0.05$), spacious (80%, $p < 0.05$), lively (80%, $p < 0.05$), safe (90%, $p < 0.01$), uniform (80%, $p < 0.05$) and bright (100%, $p < 0.01$), compared to the C2 condition. C2 was considered interesting by 80% ($p < 0.05$) of the subjects, when compared to C1. In the comparison A1 and C1, the condition A1 was judged as pleasant and warm by 100% of the subjects ($p < 0.01$), cosy by 90% ($p < 0.01$), sociable, interesting and uniform by 80% of the respondents ($p < 0.05$).

4.3.3.4 The Perception of Sociability and Warm CCT of Lighting

Participants were able to discriminate between the two different CCTs of white lighting (3000 K and 6000 K) in the outdoor environment. The luminous scenario, with warm white lighting at 3000 K (A1), was perceived as warm (by 100% of the respondents of all the studies), comfortable, and suitable for socialization by all the participants (Fig. 4.6). The 3000 K LEDs seemed to be more human-oriented, both for their chromatic aspects and also because they were perceived as less bright in comparison to the cold lighting condition (A2) and the neutral CCT (B1). Qualitative insights derived from the interviews explained that warm lighting was considered more acceptable, both in relation to previous experiences and for the contextual features. The warm tonality of light, which is perceived as "*yellow–red*", elicited a more sociable and human-oriented atmosphere. It was associated with "*warm feelings*", it "*attracts more*" and "*gives you a welcome*", and it was "*more interesting for socializing*" because "*It makes you stop maybe and stay more*".

The reference to the orange-yellowish colour of the traditional street lighting (LPS–HPS), which were located in the surrounding streets, was also addressed as a reason of selection: "*What I like is that the colour matches with the outside colour of lighting, more orange. I feel more comfortable*". In addition to this, the connection to the warm domestic "*incandescent lighting*", with "*a feel-at-home touch*", influenced the preference of the warm lighting condition, which was perceived as pleasant by 90%, 70% and 100% of respondents in the three studies (Fig. 4.6). More than this, the warm CCT defined an impression of "*warming up the space*", enhancing the thermal comfort in the cold and snowy winter. Differently, cold white lighting was

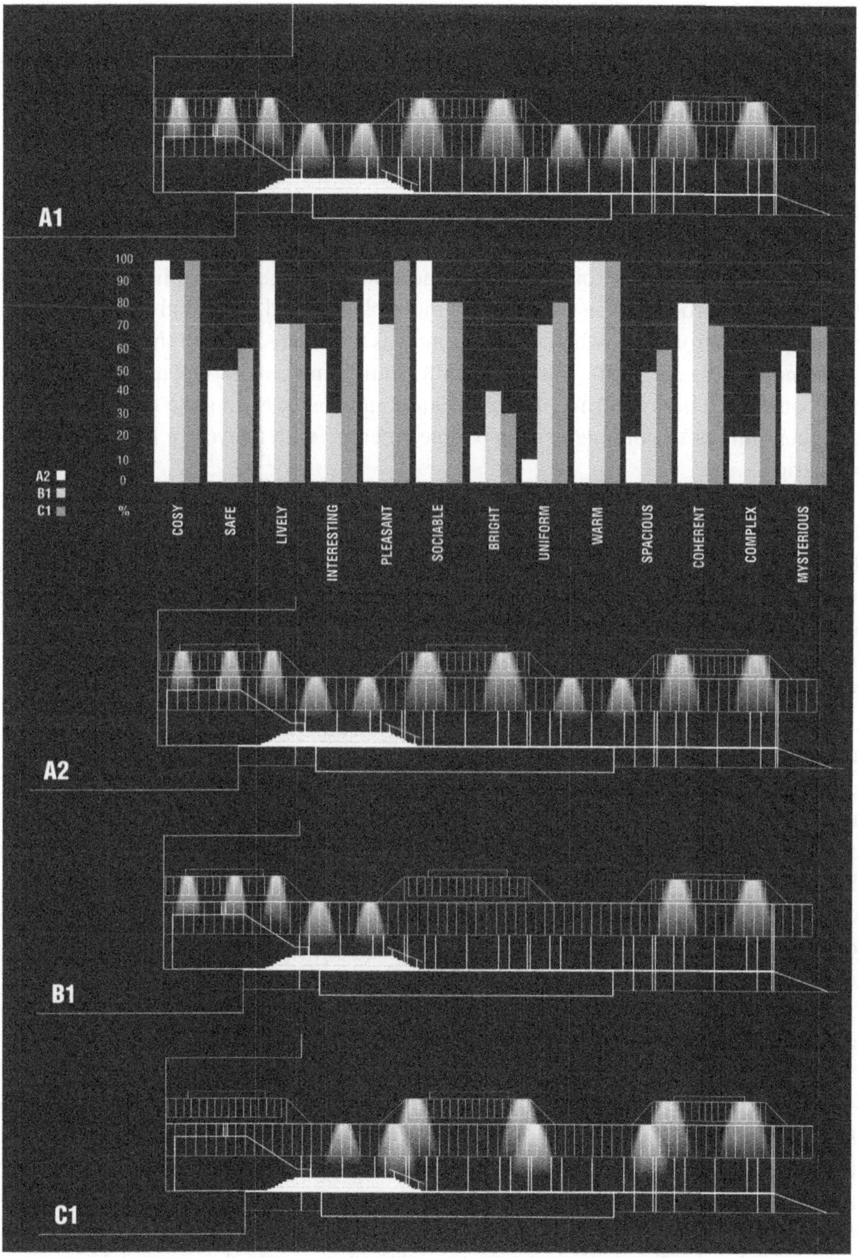

Fig. 4.6 Percentage of selection of the lighting condition A1 versus A2, B1 and C1, represented in a longitudinal section

described as "*bluer*" and "*too white*". Beside the warm uniform lighting condition (A1), the scenario considered most sociable was C1, compared to the other lighting conditions; C1 was considered safe, lively, bright, uniform and spacious compared to other uneven (B1) and layered lighting conditions (C2) (Fig. 4.7).

4.3.3.5 Lighting Distribution: Functional Versus Evocative

When comparing the uniform lighting condition A1 to the layered lighting scenario C1, 100% of participants in study 3 were more pleased and interested by the homogeneously lit environment, where the lighting was considered able to show the space in its completeness and uniformity. The uniform lighting condition was also preferred and considered best in terms of socialization and cosiness. In spite of these results, a higher number of comments from people emerged with uneven or layered lighting scenarios.

Despite safety achieving few statistically significant results, during the interviews it was a recurrent topic. There were many statements about "*more light*" as the best solution for "*safety reasons*" and darkness being associated with the denial of visibility, resulting in low safety perception and poor accessibility due to the creation of dark corners "*where people and danger can hide*". The uniform lighting condition, instead, enabled participants to observe the entire space, highlighting the preference "*to be in the light and not in the darkness*".

On the other hand, several participants showed the opposite opinion, arguing that too much light was not always positive both for people and for space, and that darkness had its own positive features, namely hiding less-important features of the space, creating a "*mysterious and more interesting*" environment, and also defining a better space in terms of socialization. In particular, a darker space with enough lighting for perceiving the shape but enough darkness to feel it was comfortable and cosy was felt to be more evocative in terms of human-scale lighting: "*It reminds me of places where you go for clubbing. Dark spots make me think about cosier meetings in the space. I feel that people in the dark talk more*".

In the layered approach (C1), the dim ambient lighting, coupled with luminous accents on visual cues, made respondents perceive the space as "*more attractive*", "*accessible*",*inviting*", "*welcoming and socially inclusive*", and more "*suitable in supporting conversations and social interaction*". C1 "*is something in between the having too much lighting and not having light*". The presence of a uniform low level of light contributed to the perception of safety and augmented visual clarity; meanwhile, luminance contrast had an emotional influence. "*It is lit and not lit. There is light and also darkness. So it makes the space more intimate*". Some participants perceived that the distribution of lighting and shadows could shape people's positions in the space: "*you can focus on people and decide to be in the darkness or in the light*".

More than this, dark and dim scenarios were clearly associated with less consumption: the evaluation of energy efficiency was one of the recurrent topics during the interviews. Even though none of the lighting settings were effectively more or

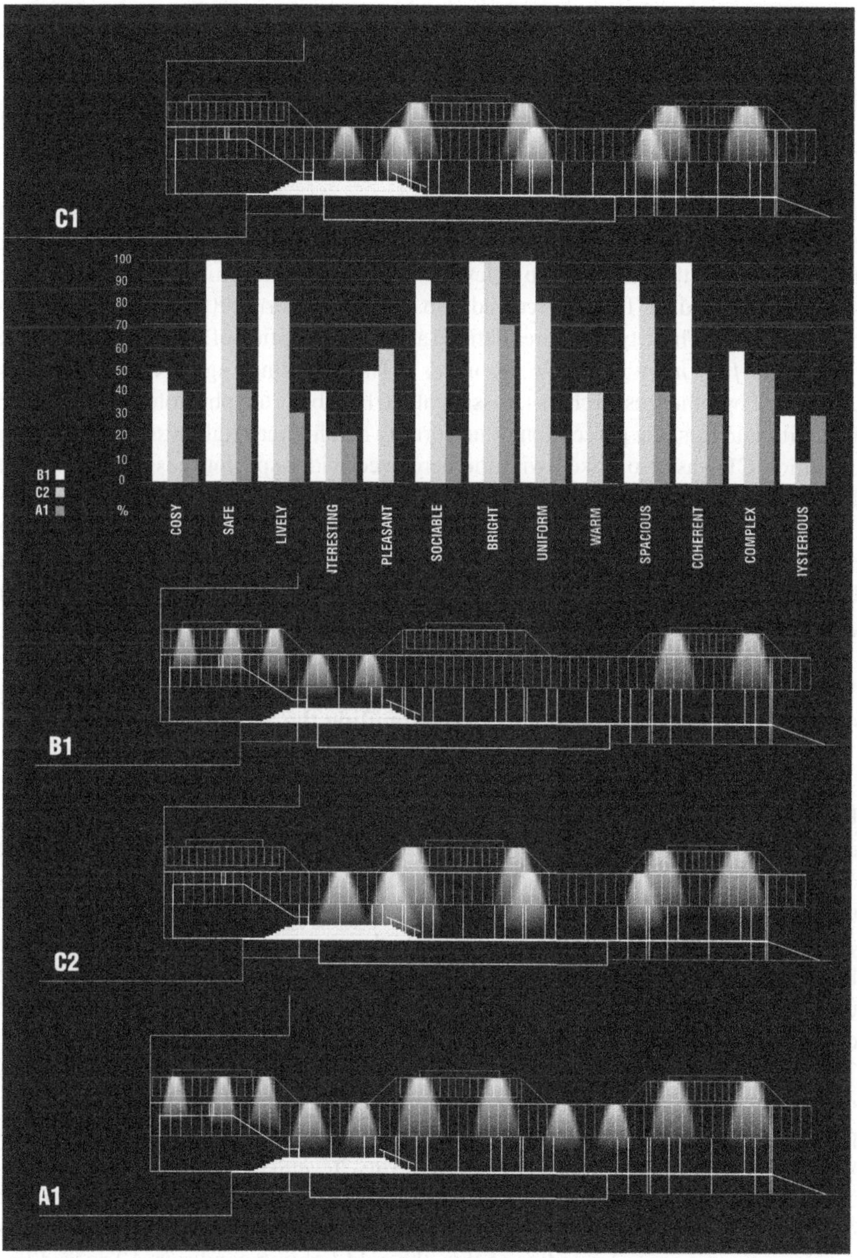

Fig. 4.7 Percentage of selection of the lighting condition C1 versus B1, C2 and A1, represented in a longitudinal section

less impactful in terms of energy consumption, people were clearly associating the perceived-less-bright environment with less consumption.

4.3.3.6 The Spatial Perception: Brightness and Uniformity

In the comparison between uniformly lit scenarios (A1 and A2), the cold CCT at 6000 K determined a higher perception of brightness (80% of the participants in the first study), even though the lighting levels of horizontal and vertical illuminance were set as equal. Cold CCT was perceived also as more uniform (90%) in comparison to the warm CCT. These two features defined also the perception of higher spaciousness: "*the space [...] more big, let's say wider*". These findings agree with the studies completed by other researchers (Rossi et al. 2012) about the physiology of the eyes. In addition to this, the perception of uniformity and higher brightness of the lighting condition C1 was also linked with the perceived wider spaciousness of the Market Hall.

In the qualitative comments, perceived higher spaciousness was also linked to darkness: the absence of perceivable boundaries contributed to define a wider, without limits and not measurable space, as also reported by previous studies in outdoor environments (Wänström Lindh 2012).

4.3.3.7 The Spatial Perception: Understanding the Space Through Its Visual Cues

Results show that lighting could change the spatial perception in terms of legibility, complexity and coherence. According to Kaplan and Kaplan (1989), the perception of the space can be predicted through the preference matrix based on these evaluative descriptors. In particular, complexity represents the richness of different visual elements in the scene, defining how intricate the scene is and how much there is to look at. Coherency is the orderly structure that helps to organize, make sense of and understand the environment. These descriptors emphasize the content of information of the environment in an ecological sense.

A uniform, warm carpet of light was perceived as highly coherent with the environment but not complex—clear, but at the same time, expected. Cold lighting (A2) was perceived as consistent with street lighting but absolutely "*not coherent for a place where people were meant to be*", because it was "*technical*", "*industrial*", "*sterile*" and "*clean*". Cold CCT has been associated with an uncomfortable feeling: "*It is too bright and not comfortable. With this, I mean that it can cause you headaches. It is less easy for the eye*". According to this, the cold lighting condition A2 scored low coherence and high complexity compared to the warm CCT lighting condition (A1). This means the environment was perceived as too visually messy to be accessed in a comfortable way.

The lighting condition C1 was perceived as highly coherent in comparison to B1 (100% of participants in study 2) in evidencing more elements of the space and in contributing to the architectural and spatial-making sense. Indeed, lighting was seen as a valuable element for revealing details of the space and changing its atmosphere and perception, enhancing the visual complexity and communicating different meanings to people. Lighting in the peripheral background was perceived as focusing on outdoor spaces; lighting in the inside of the Market Hall was perceived as enabling to focus the attention on the internal space. Lighting the elements of the space like pillars, corridors, the gates and the benches was found to be positive in defining the space and guiding the eyes in the architectural understanding.

4.4 Discussion and Insights

This research aimed to underline the importance and influence of lighting parameters on the perception of the outdoor luminous environmental atmosphere as well as on social and spatial appraisal. More than solely a stimulus–response for vision, lighting may send visual messages that may involve thinking (cognitive process) and emotions (affective process), in addition to motivation and purposeful behaviours. The scope was to prove primarily the influence of lighting design on human psychology in terms of psycho-social affective and cognitive effects of two lighting parameters: CCT and lighting distribution. From the experiments, it is evident that lighting was not the determining factor to make a place more social or sociable, but it could be one of a set of factors complementing the urban outdoor environment's social quality. Lighting was found to add meaning to a space by sending visual messages in an affective, cognitive and associative way (de Kort 2019; de Kort and Veitch 2014; "Light + Behavior Symposium—Illuminating Engineering Society" 2014). In particular, certain lighting variables seemed to contribute to a more human- and social-oriented experience of the night-time urban environment in terms of comfort and sociability, interest and coherence. The main insights related to lighting variables that make the space more human-oriented are as follows:

- warm CCT of white lighting,
- calibrated layered distribution of lighting and
- balanced luminance contrast ratio of lighting and darkness on visual cues.

Stating the exploratory nature of the research, we claim that the achieved results indicate some main suggestions for lighting designers: to make a cosier and human-scale square, warm lighting is more indicated; to make a safer space, lighting should show all the corners and possible threats (with a uniform lighting or a layered approach); to make it more interesting, lighting should highlight meaningful elements; to define a more sociable, intimate space, a juxtaposition of light and dark areas should also be arranged.

Warm lighting was perceived as cosier, as more comfortable and sociable, and as having a more human-scale tonality of lighting due to its appearance and a lower perception of brightness, and also because of contextual reference or based on culturally accepted lighting. Cold white lighting was absolutely not intended for sociability. These findings are strictly correlated to the context (location, size, shape, proximity to other features, weather conditions, presence/absence of people in the space during the experiment) and might not be extended to other situations. Other than this, the influence of the warm CCT on sociability agrees with previous findings in interior spaces (Baron et al. 1992; Heijboer 2013; Kombeiz 2017).

Uniform and not uniform lighting distributions determined two opposite impressions. One impression was the feeling of safety and functionality of the environment, which was correlated to the uniformly lit space with high coherence to expectations and low complexity in spatial decoding. These findings agree partially with the prospect-refuge theory (Fisher and Nasar 1992; Nasar and Fisher 1993) and with the results of other studies of lit squares and pedestrian areas (Calvillo Cortés and Falcón Morales 2016; Nasar and Bokharaei 2016, 2017): dark, unlit corners give the impression that they may hide information and threats, thus causing the impossibility to predict what could happen, causing fear. It has to be stated that this space was a protected square inside the university campus and this might have defined lower uncertainty and a higher safety perception. A wider understanding on the issue should focus on different environments such as open and unprotected squares to assess the lighting effects in different situations.

Uniform dim lighting combined with accent lighting on specific visual cues elicited a more mysterious and sociable space. Due to the fact that the sense of sight is contrast sensitive, the brightness contrast determines the richer emotional impact on the visual experience. In this experiment, the brightness contrast ratio depended on the following:

- the relationship of lit surfaces perceived as bright to those left in the dark,
- the luminance contrast ratio versus uniformity,
- the focus on the foreground or background and
- the focus on details in the visual environment.

Through a comparison of the results of the two phases of the studies, we could differentiate between the affective quality attributed to the place and the luminous atmosphere altered and evoked by the lighting. The atmosphere of the space was perceived as evidently enhanced by the luminous composition, which was coherent with the architecture (layered approach). These lighting conditions were able to influence the human emotional and cognitive responses (Boyce 2004) in perceiving

- Identification of visual boundaries
- Understanding of functional details and visual hierarchy
- Socially including or excluding luminous atmospheres
- Intimacy and sociability.

The interviews revealed that the different lighting conditions were perceived as positively transforming the atmosphere of the space. This is also indicating that experiments similar to this one can be effective tools to test in advance the lighting design of a space for deeper inclusiveness of the point of view of people prior to the final installation. The results of this research could provide practical design guidance for architects and interior and lighting designers, helping them to develop new lighting design solutions that take into account the psycho-social effects of lighting. In particular, the combination of functional and more evocative lighting could be suggested as a viable solution to define more human- and social-scale outdoor spaces. Evocative lighting becomes an activator of the space in terms of creating more elements to look at, sending additional meaningful information to the users, and defining an intriguing scenario that reveals features and different readings of the space (Berlyne 1954). The uniform lighting approach is coherent with the space and functional for safety issues. The warm CCT and the layered approach are more evocative for social use and more meaningful for the architecture; it is perceived as an opportunity to enhance the features of a space, even if quite anonymous and underused, and to create socially pleasant night-time spaces. In addition to this, the recurrent interest in energetic and environmental issues within an effort to use less electricity while maintaining a high-quality lit environment was perceived as important. Only by understanding the perceptual dimensions of lightness and darkness could the space accomplish functionality, evocativeness and sustainability.

Even though the participants in the studies represent the real users of the space (college students and staff), these conclusions are limited by the fact that we cannot allow for the generalisation of the gathered insights into the overall population in different age ranges: the results only refer to a population between the ages of 18 and 39 years old. As a future research, participants from a broader age sample should be analysed to understand if personal factors affect the impression of social appraisal and sociability. In addition to this, further studies might go beyond the socio-psycho affective impressions and also analyse other features such as prospect-refuge theory, legibility and wayfinding. They could also investigate more deeply the sociability in motivational terms by exploring behavioural intentions.

References

Augé M (1992) Non-lieux. Éd. du Seuil, Paris

Baron R, Rea M, Daniels S (1992) Effects of indoor lighting (illuminance and spectral distribution) on the performance of cognitive tasks and interpersonal behaviors: the potential mediating role of positive affect. Motivation and Emotion 16(1):1–33. https://doi.org/10.1007/bf00996485

Berlyne D (1954) A theory of human curiosity. Br J Psychol Gener Sect 45(3):180–191. https://doi.org/10.1111/j.2044-8295.1954.tb01243.x

Boyce P (2004) Lighting research for interiors: the beginning of the end or the end of the beginning. Light Res Technol 36(4):283–293. https://doi.org/10.1191/1477153504li118oa

BSI 5489.1:2003 + A2: 2008—Code of practice for the design of road lighting—lighting of roads and public amenity areas
Calvillo Cortés A, Falcón Morales L (2016) Emotions and the urban lighting environment. SAGE Open 6(1):215824401662970. https://doi.org/10.1177/2158244016629708
Canter D (1977) The psychology of place, 1st edn. The Architectural Press, London
Decuypere J, Capron J, Renglet M (2009) Influence of mesopic lighting conditions on Pedestrian visual Field in urban environment. In Lux Europa 2009—11th European lighting conference, pp 1177–1182. Istanbul
de Kort Y, Veitch J (2014) From blind spot into the spotlight. J Environ Psychol 39:1–4. https://doi.org/10.1016/j.jenvp.2014.06.005
de Kort Y (2019) Tutorial: theoretical considerations when planning research on human factors in lighting. LEUKOS, 1–12. https://doi.org/10.1080/15502724.2018.1558065
Den Ouden E, Haans A, van Essen H, Ross P, de Kort Y, Brankaert R (2019) Achievements and learnings of researching intelligent lighting solutions in living labs at the TU/e Intelligent Lighting Institute 10 YEARS LIVING LIGHT LABS. Eindhoven University of Technology
DiLaura D, Houser K, Mistrick R, Steffy G (2011) The lighting handbook, 10th edn. Illuminating Engineering Society
Ector Hoogstad Architecten (2012) MetaForum TU/e. Nieuwe universiteitsbibliotheek en faculteit Wiskunde & Informatica in en op historische machinehal. Rotterdam
Fisher B, Nasar J (1992) Fear of crime in relation to three exterior site features. Environ Behav 24(1):35–65. https://doi.org/10.1177/0013916592241002
HDRscope (2019) Retrieved from https://courses.washington.edu/hdrscope
Heijboer M (2013) Seeing things in a positive light. Investigating the effects of color temperature of light on sociality and de-escalation (Master of Science in Human Technology Interaction). Department of Industrial Engineering & Innovation Sciences, Eindhoven University of Technology
Hillier B, Hanson J (1984) The social logic of space. University Press, Cambridge
Inanici M (2006) Evaluation of high dynamic range photography as a luminance data acquisition system. Light Res Technol 38(2):123–134. https://doi.org/10.1191/1365782806li164oa
Kaplan R, Kaplan S (1989) The experience of nature: a psychological perspective. Cambridge University Press, New York
Kelly R (1952) Lighting as an integral part of architecture. Coll Art J 12(1):24. https://doi.org/10.2307/773361
Kombeiz O (2017) Turning the spotlight on the role of light and colors in offices: how are performance, social interactions, and social perception affected? (Ph.D. Dissertation). University of Hohenheim
Kumaragurubaran V, Inanici M (2013) Hdrscope: high dynamic range image processing toolkit for lighting simulations and analysis. In: International Building Performance Simulation Association (IBPSA). Chambery.
Light + Behavior Symposium—Illuminating Engineering Society (2014) https://www.ies.org/research/events/research-symposia/light-behavior-symposium/
Luminance HDR (2019) Retrieved from https://sourceforge.net/projects/qtpfsgui/
Nasar J, Fisher B (1993) 'Hot spots' of fear and crime: a multi-method investigation. J Environ Psychol 13(3):187–206. https://doi.org/10.1016/s0272-4944(05)80173-2
Nasar J, Bokharaei S (2016) Impressions of lighting in public squares after dark. Environ Behav 49(3):227–254. https://doi.org/10.1177/0013916515626546
Nasar J, Bokharaei S (2017) Lighting modes and their effects on impressions of public squares. J Environ Psychol 49:96–105. https://doi.org/10.1016/j.jenvp.2016.12.007
Rossi L, Zegna L, Iacomussi P, Rossi G (2012) Pupil size under different lighting sources. In: CIE 2012 'Lighting quality and energy efficiency'
Software | Space Syntax Network (2019) Retrieved from https://www.spacesyntax.net/software/

Vogel I (2008) Atmosphere metrics: a tool to quantify perceived atmosphere. In: Probing experience. From assessment of user emotions and behaviour to development of products, 1st edn. Springer Netherlands

Wänström Lindh U (2012) Understanding the Space: how distribution of light influences spatiality. In: 10th European Academy of Design conference—crafting the future

Chapter 5
A Virtual Experience of the Human-Scale Urban Lightscape

Abstract This chapter presents research results related to the influence of lighting on impressions of public outdoor urban streets using a virtual environment. Lighting parameters such as uniformity (non-uniform, uniform) and distribution (horizontal, horizontal and vertical, and multi-layered approach with horizontal, vertical, and accent lighting) were manipulated to understand their influence on the impression and evaluations of a space's accessibility, sociability, hospitality, and comfort. Three different virtual urban streets with different features (with respect to dimensions and the presence of natural elements) were considered and investigated. The research sample included 105 respondents that judged the perceived spaciousness/confinement, sociability/unsociability, hospitality/inhospitality, and privacy/publicness of each of 18 digital images of the nocturnal streets. The results showed that judgments of spaciousness, hospitality, and sociability increased with uniform, horizontal/vertical, and also horizontal/vertical/accent lighting. Conversely, privacy increased with non-uniform lighting. The emerging preliminary insights of this study inform design practices to develop hospitable and social experiences in public spaces at night.

Keywords Urban design · Lighting design · Psychology · Aesthetics · Meaning · Preference · Experiment · Public space · Human-scale · Sociability

5.1 Introduction

Atmospheric elements can alter and guide the perceptions and impressions of a space. However, we do not know whether lighting design has a strong effect on the impression of a given context. This research tested whether and how different lighting variables and conditions affected judgments of spaciousness, publicness, sociability, and hospitality in public streets after dark. We chose to study those impressions because they may affect either the assessment or activity level or both, and attract more people and increase social interactions thereby.

Upgrading lighting fixtures for outdoor spaces with LEDs has increased the ability to vary the light in a space with respect to intensity, chromaticity, and distribution

D. Casciani, *The Human and Social Dimension of Urban Lightscapes*,
PoliMI SpringerBriefs, https://doi.org/10.1007/978-3-030-57165-8_5

to obtain the lighting design desired. Conversely, the endless possibilities LED technologies allow require a solid understanding of the way lighting influences people's impressions of a space. This is pivotal to use the technology as a service for users of urban public spaces to create human-centred, tailored lighting.

This research focused on two primary lighting modes to assess the responses to lighting in public streets after dark: Non-uniform/uniform and three different kind of lighting distributions (horizontal, vertical, accent). Research has suggested that lighting distribution is a salient feature that influences human's perceptions of interior spaces (Hawkes et al. 1979; Veitch and Newsham 1998; Durak et al. 2007; Flynn 1988; Küller et al. 2006; Miwa and Hanyu 2006; Barazawa and Hanyu 2013).

To understand the effects of lighting in public squares at night, one must study people's responses to those settings. The research focused on outdoor environments, and tested hypotheses based on indoor studies. Given that human behaviours and responses to lighting differ across settings that are different in purpose (Boyce et al. 2000; Butler and Biner 1987), we did not expect to find the same results in private indoor settings and public streets. Theory and research suggested three sets of hypotheses for the lighting conditions:

- **Hypothesis 1 (H1)**: Uniform lighting gives the perception of greater spaciousness, publicness, sociability, and hospitality than does non-uniform lighting. Consequently, uniform lighting also increases the behavioural intention to use the space socially.
- **Hypothesis 2 (H2)**: Horizontal/vertical (HV) lighting gives the perception of greater spaciousness, publicness, sociability, and hospitality than does horizontal (H) lighting. Consequently, HV lighting also increases the behavioural intention to use the space socially.
- **Hypothesis 3 (H3)**: A horizontal/vertical distribution with accent (HVA) lighting gives the perception of greater sociability and hospitality than does horizontal (H) lighting. Hence, HVA lighting also increases the behavioural intention to use the space socially.

5.2 Methodology

5.2.1 Participants

The researcher recruited a convenience sample of volunteers through online requests. To reduce respondents' effort, each subject was assigned to only one case study. The case studies were presented randomly. The sample included 105 respondents, divided as follows: 33 participants (31.43%, 21 females, 12 males) for the first case study, "Car Promenade," 39 (37.14%, 20 females, 19 males) for the second, "Residential Promenade," and 33 (31.43%, 21 females, 12 males), for the third, "Urban Garden Promenade." 59.05% of the sample was female and 40.95% was male. The mean age was 35.29 ± 8.31 years. 78.10% of the respondents were Italian and this percentage was homogeneous among groups too. The respondents' educational background was largely Master's degree (57.14%), Bachelor's degree (16.19%), Ph.D. (15.24%), and diploma (11.43%), which was distributed similarly in the three groups. With respect to previous knowledge of lighting design, 69.52% of the sample were novices, and 30.48% were experts. The sample size was determined using a power analysis performed with G*Power ($p = 0.05$, effect size = 0.50, power = 0.80).

5.2.2 Lighting Stimuli: Design and Description

The research tested the mode of lighting separately in three simulated streets that varied in area, the presence of nature in the surroundings, and the type of lamp and light fixture used. Street A, "Car Promenade," was 18 m large in total size with 2 lateral walkaways 4 m wide. The street represented an urban environment surrounded by high buildings with few green elements along the street. Street B, "Residential Promenade," was a 20 m wide pedestrian path with a surrounding of trees and smaller residential buildings. Street C, "Urban Garden Promenade," was a 16 m wide built environment with two lateral spaces for plants and greenery in the surroundings and a central pedestrian passage of approximately 8 m. For each street, other factors, such as the colour of the light, shadows, and the number and locations of people, were kept constant. For realism and scale, people were placed in lighted areas to avoid affecting the perception of security and safety. Moreover, the positioning, height, and field of view were set in the same way to provide the same perspectival view in the simulated and rendered images.

The studies used 3Ds Max with the rendering engine Mental Ray to create 18 color slides of the streets that varied in the following modes of lighting: Non uniform versus uniform; horizontal versus horizontal; vertical versus horizontal, and vertical with accent lighting (Tables 5.1 , 5.2 and 5.3). All of the lighting fixtures used for the simulation were equipped with LED lighting, which have good efficiency, color rendering, and long life, and are used increasingly in public places. The difference between uniform/non-uniform was achieved by increasing the distances between the poles. The difference between the lighting distribution was achieved by using façade lighting to light the horizontal surfaces and accent lighting to light the greenery and benches both directly and indirectly.

Crossing each mode with each other mode resulted in six mixes of lighting modes for each street (Car Promenade, Residential promenade, Urban Garden Promenade): Non-uniform horizontal lighting, non-uniform horizontal and vertical lighting, non-uniform horizontal, vertical with accent lighting, uniform horizontal lighting, uniform horizontal and vertical lighting, uniform horizontal, and vertical with accent lighting.

Each image was shown in colour (at a 300-dpi resolution) on a grey background through a digital survey presented using SurveyGizmo. The survey did not use a timing function. Participants could respond at their own pace and were not told how long to view each slide.

Table 5.1 Street A, "Car Promenade"

	Uniform	Non-uniform
Horizontal		
Horizontal and vertical		
Horizontal, Vertical, accents		

Table 5.2 Street B, "Residential Promenade"

	Uniform	Non-uniform
Horizontal		
Horizontal and vertical		
Horizontal, Vertical, accents		

Table 5.3 Street C, "Urban Garden Promenade"

	Uniform	Non-uniform
Horizontal		
Horizontal and vertical		
Horizontal, Vertical, accents		

5.2.3 Instruments and Procedure

To assess perceived spaciousness/confinement, sociability/unsociability, hospitality/inhospitality and privacy/publicness, the research used 7-point bipolar items: spacious-confined, sociable-unsociable, public–private, hospitable/inhospitable, and uniform-non-uniform. These environmental descriptors were adopted from previous studies that were designed to assess spaciousness and privacy in interior spaces (Flynn 1988). The participants read one of the following items beneath each image.

For perceived spaciousness/confinement, the text stated, "How spacious is this place for you?" Participants could check one of the following answers: Very strongly spacious, strongly spacious, spacious, neither spacious nor confined, confined, strongly confined, very strongly confined.

For perceived sociability/unsociability, the text read, "How sociable is this place for you?" and was accompanied by the following possible answers: Very strongly sociable, strongly sociable, sociable, neither sociable nor unsociable, unsociable, strongly unsociable, very strongly unsociable.

For perceived privacy/publicness, the text read, "How public is this place for you?" The participants could check one of the following answers: Very strongly public, strongly public, public, neither public nor private, private, strongly private, very strongly private.

For perceived accessibility/inaccessibility, the text stated, "How hospitable is this place for you?" and was accompanied by the following possible answers: Very strongly hospitable, strongly hospitable, hospitable, neither hospitable nor inhospitable, inhospitable, strongly inhospitable, very strongly inhospitable.

For perceived uniformity/non-uniformity, the text read, "How uniform is the lighting in this place for you?" The participants could check one of the following answers: Very strongly uniform, strongly uniform, uniform, neither uniform nor non-uniform, non-uniform, strongly non-uniform, very strongly non-uniform.

The survey also had three statements about the behavioural intention to use the space socially: "I would like to sit in this place," "I would like to meet a person in this place," "I would like to invite someone to go out with me in this place." For each, the text below each slide asked participants how much the statement applied to their likely behaviour in relation to the place (from 1, very strongly agree, to 7, very strongly disagree).

The order of the 48 questions (eight judgments for each of the 6 streets for each study) was randomized across participants, so that each participant saw the questions in a different order.

Before the ratings, participants answered questions about the categories, their gender, age, and knowledge of the lighting design discipline (expert/novice).

5.3 Data Analysis and Results

5.3.1 *Statistical Analysis Methodology*

Data collected by SurveyGizmo were analysed using STATA16.

Descriptive statistics were performed using means and standard deviations for quantitative data, and frequencies and percentages for qualitative data.

Given the sample size, the Shapiro–Wilk's test was performed to assess whether the variables were distributed normally. Tables 5.4 and 5.5 show the means and standard deviations for each of the 18 conditions. A repeated-measures ANOVA was conducted separately for the perceived spaciousness/confinement, sociability/unsociability, hospitality/inhospitality and privacy/publicness, together with the behavioural intention to use the space socially ("like to sit," "invite", "meet a person") and the two different main lighting conditions: Uniform (U)/Non-Uniform (NU), Horizontal (H)/Horizontal and Vertical (HV)/Horizontal and Vertical with Accents (HVA). We used a paired *t*-test for comparison and Cohen's *d* paired samples to evaluate the effect size. Statistical significance was set at $p < 0.05$.

5.3.1.1 Lighting Distribution: Non-uniform (NU)/Uniform (U)

The participants rated the uniform (U) lighting conditions significantly higher for the perception of sociability and hospitality in all three streets (A—Car Promenade, B—Urban Residential, C—Urban Garden) ($p < 0.001$). The Cohen's *d*s in the last column of the table indicated a small-to-medium effect size.

For the perception of spaciousness, the respondents assigned significantly lower scores to the non-uniform (NU) than the uniform (U) lighting distribution in all three streets: $p < 0.01$ for A—Car Promenade, $p < 0.001$ for C—Urban Garden, and $p < 0.05$ for B—Urban Residential. The Cohen's *d*s in the last column of the table indicated a small effect size.

A similar trend was observed for the perception of publicness: A—Car promenade and C—Urban Garden were considered more public with U lighting ($p < 0.001$), while there was no statistically significant difference in the case of B—Urban Residential. The Cohen's *d*s in the last column of the table indicated a small-to-medium size effect.

The behavioural intention to use the space socially ("like to sit") received significantly higher scores with the uniform than the non-uniform lighting conditions in all three streets ($p < 0.001$) with a small effect size.

A similar trend was found in the behavioural intention to invite someone to the space: U lighting conditions were preferred significantly over the NU conditions ($p < 0.001$ for street B and $p < 0.01$ for streets A and C. The Cohen's *d*s indicated a small-sized effect.

Table 5.4 Means and standard deviations for each of the 18 lighting conditions with the dependent variables of atmospheric attributes

		Spaciousness/confinement	Privacy/publicness	Hospitality/inhospitality	Sociability/unsociability	Uniformity/non-uniformity
Street	Lighting conditions	M (SD)	M (SD)	M (SD)	M (SD)	M (SD)
A	NUH	3.79 (±1.34)	4.55 (±1.25)	2.61 (±1.20)	2.58 (±1.30)	2.85 (±1.33)
	NUHV	4.42 (±0.90)	4.70 (±0.98)	3.79 (±0.82)	3.81 (±0.92)	3.42 (±0.94)
	NUHVA	4.79 (±0.96)	4.97 (±0.88)	4.39 (±0.97)	4.45 (±0.79)	3.64 (±1.25)
	UH	4.15 (±1.23)	4.76 (±1.15)	3.36 (±0.90)	3.36 (±1.11)	3.55 (±1.00)
	UHV	4.88 (±1.14)	5.18 (±0.85)	4.45 (±1.15)	4.39 (±1.20)	4.85 (±1.15)
	UHVA	5.09 (±0.95)	5.30 (±0.85)	4.79 (±0.96)	4.79 (±0.99)	4.82 (±1.04)
B	NUH	4.26 (±1.14)	4.59 (±1.02)	3.46 (±1.02)	3.59 (±0.97)	3.10 (±1.25)
	NUHV	4.64 (±0.93)	4.92 (±0.87)	4.18 (±0.94)	4.13 (±0.83)	3.33 (±1.06)
	NUHVA	4.95 (±0.97)	5.0 (±0.56)	4.38 (±1.09)	4.51 (±1.17)	3.72 (±1.32)
	UH	4.67 (±0.84)	4.76 (±1.15)	4.08 (±0.74)	4.26 (±0.79)	4.0 (±1.27)
	UHV	4.84 (±0.93)	5.03 (±0.67)	4.36 (±0.93)	4.54 (±0.82)	4.46 (±1.12)
	UHVA	5.0 (±1.03)	5.33 (±0.93)	4.85 (±1.14)	4.92 (±0.90)	4.62 (±1.27)
C	NUH	4.06 (±1.06)	4.67 (±0.69)	3.94 (±0.97)	3.81 (±0.92)	3.79 (±1.02)
	NUHV	4.27 (±0.88)	4.60 (±0.86)	4.21 (±0.82)	4.24 (±0.83)	3.64 (±0.90)
	NUHVA	4.51 (±0.91)	5.0 (±0.56)	4.88 (±0.65)	4.82 (±0.73)	4.12 (±1.05)
	UH	4.33 (±1.08)	4.88 (±0.99)	4.36 (±1.06)	4.39 (±1.09)	4.52 (±1.03)
	UHV	4.97 (±1.19)	5.33 (±0.65)	5.0 (±0.94)	5.06 (±0.83)	5.09 (±0.91)
	UHVA	4.94 (±1.09)	5.39 (±0.70)	5.27 (±0.88)	5.21 (±0.70)	5.12 (±1.02)
M		4.59 (±1.03)	4.94 (±0.87)	4.24 (±0.95)	4.27 (±0.94)	4.04 (±1.11)

Table 5.5 Means and standard deviations for each of the 18 lighting conditions with the dependent variables of behavioural intentions

		Like to sit	Meet a person	Invite someone
Street	Lighting conditions	M (SD)	M (SD)	M (SD)
A	NUH	2.48 (±0.94)	2.27 (±0.80)	2.73 (±1.40)
	NUHV	3.70 (±0.95)	3.70 (±0.95)	3.58 (±1.09)
	NUHVA	4.64 (±1.11)	4.42 (±1.09)	4.30 (±1.13)
	UH	3.39 (±1.14)	3.33 (±1.19)	3.09 (±1.10)
	UHV	4.42 (±1.25)	4.48 (±1.28)	4.15 (±1.50)
	UHVA	4.82 (±1.18)	4.58 (±1.17)	4.45 (±1.33)
B	NUH	3.28 (±1.15)	3.36 (±1.40)	3.21 (±1.38)
	NUHV	3.97 (±1.18)	3.69 (±1.03)	3.44 (±1.05)
	NUHVA	4.41 (±1.43)	4.21 (±1.42)	4.08 (±1.42)
	UH	4.03 (±1.01)	3.85 (±1.01)	3.59 (±1.02)
	UHV	4.41 (±1.09)	4.33 (±1.08)	4.05 (±1.07)
	UHVA	4.90 (±1.19)	4.67 (±1.32)	4.62 (±1.43)
C	NUH	3.82 (±1.16)	3.70 (±1.07)	3.61 (±1.06)
	NUHV	4.09 (±0.95)	4.07 (±0.90)	4.0 (±1.12)
	NUHVA	4.82 (±0.85)	4.76 (±1.00)	4.67 (±1.19)
	UH	4.27 (±1.15)	4.18 (±1.36)	3.97 (±1.21)
	UHV	4.73 (±1.04)	4.85 (±0.97)	4.67 (±1.08)
	UHVA	5.0 (±0.87)	5.03 (±0.92)	4.67 (±1.14)
M		4.18 (±1.09)	3.94 (±1.11)	3.94 (±1.21)

For the behavioural intention to meet a person, U lighting was preferred significantly over NU lighting: $p < 0.001$ for the three street environments and a small-sized effect.

The perception of uniformity was greater in the U lighting conditions shown in the pictures. The Cohen's *d*s for the judgment of uniformity showed large effects compared to the other factors. Table 5.6 shows the results for non-uniform/uniform lighting in each street.

5.3.1.2 Lighting Distribution: Horizontal (H)/Horizontal and Vertical (HV)/Horizontal and Vertical with Accents (HVA)

In observing and comparing the three different lighting conditions: Horizontal (H), horizontal-vertical (HV), and horizontal-vertical-accent (HVA), a general trend emerged in all findings: the HVA lighting condition was perceived to be more spacious, sociable, public, hospitable, and uniform than were the H lighting conditions. In addition, the HVA lighting condition was also the one that was most highly

Table 5.6 Non-uniform/uniform

Rating/Street	Non-Uniform (NU) M (SD)	Uniform (U) M (SD)	t	p	d
Spacious					
A—Car Promenade	4.33 (±1.15)	4.71 (±1.17)	−3.01	0.003**	0.25
B—Urban Residential	4.62 (±1.05)	4.84 (±0.94)	−2.14	0.034*	0.18
C—Urban Garden	4.28 (±0.96)	4.75 (±1.15)	−4.81	0.000***	0.35
Public					
A—Car Promenade	4.74 (±1.06)	5.08 (±0.98)	−3.45	0.000***	0.26
B—Urban Residential	4.89 (±0.94)	5.03 (±0.82)	−1.80	ns	
C—Urban Garden	4.76 (±0.73)	5.20 (±0.82)	−5.14	0.000***	0.48
Sociable					
A—Car Promenade	3.62 (±1.28)	4.18 (±1.25)	−4.81	0.000***	0.35
B—Urban Residential	4.08 (±1.06)	4.57 (±0.87)	−5.12	0.000***	0.41
C—Urban Garden	4.29 (±0.92)	4.89 (±0.94)	−5.87	0.000***	0.55
Hospitable					
A—Car Promenade	3.60 (±1.24)	4.20 (±1.17)	−5.70	0.000***	0.41
B—Urban Residential	4.01 (±1.09)	4.43 (±0.99)	−3.95	0.000***	0.32
C—Urban Garden	4.34 (±0.91)	4.88 (±1.02)	−5.54	0.000***	0.46
Uniform					
A—Car Promenade	3.30 (±1.22)	4.40 (±1.22)	−7.97	0.000***	0.83
B—Urban Residential	3.38 (±1.23)	4.36 (±1.23)	−8.89	0.000***	0.70
C—Urban Garden	3.85 (±1.00)	4.91 (±1.02)	−8.60	0.000***	1.01
Like to sit					
A—Car Promenade	3.61 (±1.33)	4.21 (±1.33)	−5.09	0.000***	0.37
B—Urban Residential	3.89 (±1.33)	4.44 (±1.15)	−4.71	0.000***	0.36
C—Urban Garden	4.24 (±1.07)	4.67 (±1.06)	−3.53	0.000***	0.31
Meet a person					
A—Car Promenade	3.46 (±1.30)	4.13 (±1.33)	−5.85	0.000***	0.41
B—Urban Residential	3.75 (±1.33)	4.28 (±1.19)	−4.19	0.000***	0.33
C—Urban Garden	4.17 (±1.08)	4.69 (±1.15)	−4.51	0.000***	0.37
Invite a person					
A—Car Promenade	3.54 (±1.37)	3.90 (±1.43)	−2.72	0.008**	0.20
B—Urban Residential	3.57 (±1.33)	4.09 (±1.25)	−3.95	0.000***	0.31
C—Urban Garden	4.09 (±1.20)	4.43 (±1.21)	−2.76	0.007**	0.22

Means and Statistics for Each Street, with *** $p < 0.001$, ** $p < 0.01$ and * $p < 0.05$ level of significance preferred with respect to social behavioural intentions ("like to sit," "meet a person," and "invite a person") over the H condition. Table 5.7 shows the results for the H/HV/HVA lighting comparison.

In particular, the three streets were perceived to be more spacious in the case of HVA lighting than H lighting ($p \leq 0.001$). The Cohen's *d*s indicated a medium size effect. This trend was also confirmed when HV and H lighting conditions were compared: The judgment of spaciousness was significantly greater for the HV than for the H lighting ($p < 0.001$ for A—Car promenade and C—Urban Garden, and $p < 0.05$ for B—Urban Residential. In this case, the Cohen's *d*s indicated a small to medium effect size. When HV and HVA lighting were compared, only one statistically significant difference emerged, the A—Car Promenade ($p < 0.05$) with a small effect size.

For the judgment of sociability, HVA versus HV lighting in the C- Urban promenade was statistically significant ($p < 0.01$). H lighting had the lowest mean value, while HVA had the highest. The judgment of sociability was higher for HV than H lighting in street A with a larger effect ($d = 0.90$) than in streets B and C. With respect to social perceptions, HVA lighting had higher scores than did H lighting with a large effect in streets A ($d = 1.62$) and C ($d = 0.94$) and a medium effect size in street B ($d = 0.71$). HVA lighting had higher scores than did HV lighting with a small to medium effect.

With respect to public perception, A—Car promenade was perceived to be more public with HV and HVA than H lighting ($p < 0.05$ for H vs. HV and $p < 0.01$ for H vs. HVA). There was no significant difference between HV and HVA in A—Car Promenade. B—Urban residential was perceived to be more public with HV and HVA than with H lighting ($p \leq 0.001$ for H vs. HV and H vs. HVA, $p < 0.05$ for HV vs. HVA). C—Urban garden was perceived to be more public with HVA lighting than H and HV lighting ($p < 0.001$ for H vs. HVA and $p < 0.05$ for HV vs. HVA). The Cohen's *d* indicated a generally small size effect.

The rating of hospitality was higher for HVA than H lighting ($p < 0.001$) with a large effect in street A ($d = 1.72$) and C ($d = 0.95$) and a medium effect size in B ($d = 0.73$). HV lighting scored higher as well with respect to hospitality than did H lighting for all three streets ($p < 0.001$) with a larger effect in street A ($d = 1.02$) than in B and C.

The three streets were perceived to be significantly more hospitable with HVA than HV lighting ($p < 0.01$ for streets A and B, $p < 0.001$ for C street with a small to medium size effect.

In relation to the perception of uniformity, HV was considered more uniform than was H lighting in A—Car Promenade ($p < 0.001$) with a medium effect size ($d = 0.67$). The other two streets did not differ significantly. HVA lighting had higher mean values than did H lighting in all three streets with a medium size effect for the A—Car promenade ($p < 0.001$), while a small size effect was found in B and C ($p < 0.01$). Uniformity was perceived to be greater with HVA than HV lighting ($p < 0.05$) only for the C—Urban garden. Cohen's *d*s indicated a small effect.

With respect to social behavioural intentions ("like to sit," "meet a person," "invite a person"), the HVA lighting condition was preferred to the other two conditions.

Table 5.7 Lighting distribution of horizontal (H)/horizontal and vertical (HV)/horizontal and vertical with accents (HVA)

Rating/street	H	HV	HVA	H versus HV			H versus HVA			HV versus HVA		
	M (SD)	M (SD)	M (SD)	t	p	d	t	p	d	t	p	d
Spaciousness												
A	3.97 (±1.29)	4.65 (±1.05)	4.94 (±0.96)	−4.11	0.000***	0.48	−5.29	0.000***	0.77	−2.49	0.016*	0.22
B	4.46 (±1.02)	4.74 (±0.93)	4.97 (±0.99)	−2.48	0.015*	0.22	−3.47	0.001***	0.42	−1.91	ns	–
C	4.2 (±1.07)	4.62 (±1.09)	4.73 (±1.02)	−3.53	0.000***	0.31	−4.54	0.000***	0.41	−0.78	ns	–
Public												
A	4.65 (±1.20)	4.94 (±0.94)	5.14 (±0.88)	−2.29	0.025*	0.2	−3.30	0.002**	0.37	−1.82	ns	–
B	4.67 (±0.89)	4.97 (±0.77)	5.24 (±0.89)	−3.57	0.000***	0.29	−4.24	0.001**	0.55	−2.40	0.019*	0.25
C	4.77 (±0.86)	4.97 (±0.84)	5.2 (±0.66)	−1.69	ns	–	−4.28	0.000***	0.46	−2.16	0.035*	0.23
Sociable												
A	2.97 (±1.26)	4.11 (±1.10)	4.62 (±0.91)	−7.24	0.000***	0.9	−10.09	0.000***	1.62	−4.02	0.000***	0.42
B	3.92 (±0.94)	4.33 (±0.85)	4.72 (±1.06)	−3.67	0.000***	0.37	−5.96	0.000***	0.71	−3.77	0.000***	0.32
C	4.11 (±1.04)	4.65 (±0.92)	5.02 (±0.73)	−4.26	0.000***	0.46	−6.77	0.000***	0.94	−3.15	0.003*	0.35
Hospitable												

(continued)

Table 5.7 (continued)

Rating/street	H	HV	HVA	H versus HV			H versus HVA			HV versus HVA		
	M (SD)	M (SD)	M (SD)	t	p	d	t	p	d	t	p	d
A	2.98 (±1.12)	4.12 (±1.05)	4.59 (±0.98)	−7.94	0.000***	1.02	−12.25	0.000***	1.72	−3.96	0.002**	0.37
B	3.77 (±0.94)	4.27 (±0.94)	4.62 (±1.13)	−4.14	0.000***	0.44	−6.69	0.000***	0.73	−2.71	0.008**	0.26
C	4.15 (±1.03)	4.61 (±0.96)	5.08 (±0.79)	−3.17	0.000***	0.37	−7.31	0.000***	0.95	−3.96	0.000***	0.44
Uniform												
A	3.2 (±1.22)	4.14 (±1.26)	4.23 (±1.29)	−5.13	0.000***	0.67	−5.06	0.000***	0.74	−0.66	ns	–
B	3.55 (±1.33)	3.9 (±1.22)	4.17 (±1.36)	−1.95	ns	–	−2.90	0.005**	0.37	−1.88	ns	–
C	4.15 (±1.08)	4.36 (±1.16)	4.62 (±1.15)	−1.67	ns	–	−3.07	0.003**	0.33	−2.01	0.049*	0.17

Means and Statistics for Each Street with *** $p < 0.001$, ** $p < 0.01$ and * $p < 0.05$ level of significance

In particular, HV was preferred over H lighting largely for the behavioural intention to sit in the space. All differences were significant: A and B ($p < 0.001$), C ($p < 0.05$). Cohen's *ds* show a large effect size in street A ($d = 0.92$) and a small effect in streets B and C ($d = 0.38, 0.25$). HVA was preferred to H lighting for the behavioural intention to sit in the space ($p < 0.001$) in all three environments. Cohen's *ds* showed a large effect in Street A ($d = 1.80$) and a medium effect in streets B and C ($d = 0.72, 0.75$). HVA was also preferred to HV lighting ($p \leq 0.001$). The Cohen's *ds* indicated a small-to-medium effect.

A similar trend was found in the behavioural intention to meet a person: HV was preferred to H lighting. The difference in street A was significant at $p < 0.001$, and at $p < 0.01$ in streets B and C. The Cohen's *ds* showed a large effect in Street A ($d = 1.09$) and a small effect in streets B and C ($d = 0.27, 0.36$). HVA was preferred to H lighting for the behavioural intention to meet a person in the three streets ($p < 0.001$). Cohen's *ds* showed a large effect in Street A ($d = 1.68$) and a medium effect in streets B ($d = 0.54$) and C ($d = 0.78$). HVA was also preferred to HV lighting ($p < 0.001$ for street A, $p < 0.01$ for streets B and C). The Cohen's *ds* indicated a small effect size.

HV was preferred to H lighting for the behavioural intention to invite a person to the street (street A, $p < 0.001$, street C, $p < 0.01$, and street B, $p < 0.05$). The Cohen's *ds* showed a medium to small effect. HVA was preferred significantly more than H lighting for this social behavioural intention ($p < 0.001$ for the three streets). Cohen's *ds* showed a large effect in Street A ($d = 1.20$) and a medium effect in streets B ($d = 0.62$) and C ($d = 0.66$). HVA was also preferred to HV lighting (street A, $p < 0.01$, street B, $p < 0.001$, street C, $p < 0.05$). The Cohen's *ds* showed a small effect (Table 5.8).

5.4 Discussion and Insights

The analyses showed that the lighting conditions affected the perceptual/cognitive impressions of the streets, as well as the respondents' behavioural intentions to use them socially. The lighting conditions affected perceptions of spaciousness and confinement, publicness and privacy, hospitality and sociability in the simulated outdoor streets. As H1 predicted, the uniform (U) lighting increased the streets' perceived spaciousness, while the non-uniform (NU) lighting increased the participants' perceived privacy, and decreased their perception of publicness. This was particularly true for streets A (Car Promenade) and C (Urban garden promenade). Similar effects were achieved in Flynn's (1988) study of office interiors and Nasar and Bokharaei's (2016) study of outdoor squares. Conversely, consistently and significantly, street B (Residential Promenade) was not found to be more spacious and public with uniform lighting. This difference may stem from the spaces' diverse settings and typology. As H1 also predicted, sociability and hospitality were perceived to be greater with uniform lighting conditions, and the participants recognized that this condition was uniform compared to the others. The behavioural intentions to use

Table 5.8 Lighting distribution of horizontal (H)/horizontal and vertical (HV)/horizontal and vertical and accents (HVA)

Rating/street	H	HV	HVA	H versus HV			H versus HVA			HV versus HVA		
	M (SD)	M (SD)	M (SD)	t	p	d	t	p	d	t	p	d
Like to sit												
A	2.94 (±1.14)	4.06 (±1.16)	4.73 (±1.14)	−7.70	0.000***	0.92	−10.00	0.000***	1.8	−4.27	0.001**	0.48
B	3.65 (±1.14)	4.19 (±1.15)	4.65 (±1.33)	−3.67	0.000***	0.38	−6.20	0.000***	0.72	−3.78	0.000***	0.29
C	4.05 (±1.17)	4.41 (±1.04)	4.91 (±0.85)	−2.44	0.017*	0.25	−5.40	0.000***	0.75	−3.65	0.001**	0.43
Meet a person												
A	2.8 (±1.14)	4.09 (±1.19)	4.5 (±1.13)	−9.34	0.000***	1.09	−10.90	0.000***	1.68	−2.76	0.007**	0.28
B	3.6 (±1.24)	4.01 (±1.10)	4.44 (±1.38)	−2.82	0.006**	0.27	−4.86	0.000***	0.54	−3.01	0.004**	0.26
C	3.94 (±1.24)	4.45 (±1.01)	4.89 (±0.96)	−3.27	0.002**	0.36	−6.23	0.000***	0.78	−3.77	0.000**	0.36
Invite a person												
A	2.91 (±1.26)	3.86 (±1.33)	4.38 (±1.22)	−4.90	0.000***	0.64	−7.21	0.000***	1.2	−3.18	0.002**	0.32
B	3.4 (±1.22)	3.74 (±1.10)	4.35 (±1.44)	−2.24	0.028*	0.23	−5.18	0.000***	0.62	−4.29	0.000***	0.38
C	3.79 (±1.20)	4.33 (±1.14)	4.67 (±1.15)	−3.53	0.008**	0.38	−5.78	0.000***	0.66	−2.37	0.021*	0.22

Means and Statistics for Each Street, with *** $p < 0.001$, ** $p < 0.01$ and * $p < 0.05$ level of significance the spaces socially by sitting on the benches, and meeting or inviting a person to the streets were enhanced with the uniform lighting, which achieved higher scores than did the non-uniform condition. These results confirmed the prospect-refuge theory (Fisher and Nasar 1992), which claims that in darkness, an environment with non-uniform lighting can be perceived as threatening and inaccessible or unsociable (Herzog and Flynn-Smith 2001; Nasar and Jones 1997). The uniform condition offered a clear view, while the non-uniform lighting increased the perceived darkness and heightened uncertainty and fear. These results could be investigated further by asking participants about their safety perceptions.

H2 was also supported in the majority of the simulated streets in this research, as variations in spatial distribution influenced the perceptual atmospheric attributes largely. The horizontal/vertical (HV) lighting distribution influenced the perception of greater spaciousness, publicness, sociability, and hospitality in the three streets more than did the horizontal (H) lighting. HV lighting also increased the behavioural intention to use the space socially more than did the H lighting in the three streets.

The horizontal/vertical with accent (HVA) lighting gave the perception of greater sociability and hospitality than did the H lighting, thus supporting H3. HVA lighting was also found to increase the behavioural intention to use the spaces socially. This confirms the research findings of Hargroves (2001) about layered lighting and overall lighting impression. People prefer lighting schemes that provide more contrast and shading through a good general uniform illuminance coupled with visual accents and decorative lighting elements to highlight certain features within the environment such as trees and seating.

In this research, we used lighting conditions at 4000 K CCT, which is used most widely in outdoor environments with LEDs to conserve energy. In other studies, dim, warm, white lighting may have elicited positive affective responses, which consequently could encourage positive interpersonal behaviours in social situations, such as evaluating an imaginary employee and enhancing collaboration (Baron et al. 1992) or limiting anti-social and aggressive behaviours, reducing social distance, and increasing the propensity for positive social behaviour (e.g., being kinder, smiling more, behaving more politely: Heijboer 2013).

This study used digital visualizations to assess the perception and evaluation of an outdoor space's luminous atmosphere. The use of controlled, digital coloured images rather than experiences in real environments has been found to be a reliable and valuable research tool in various studies (Hendrick et al. 1977; Newsham et al. 2005), and the findings have revealed that there is a high correlation between preferences for colour images and preferences on-site.

5.5 Conclusions

This research highlighted an interesting, reliable approach and tool used to explore the psychological effects of the lighting conditions on public streets. In addition, some insights and findings emerged about the lighting conditions that can help enhance the perceptions of outdoor environments' accessibility, sociability, spaciousness, and publicness. A specific lighting setting also influenced the social behavioural intentions to sit and meet in, and invite someone to, the space. Because lighting has been found to influence the perception of a space, this digital research tool and methodology could be quite helpful in generating insights into the way to proceed with an outdoor lighting design project. Lighting researchers, designers, and professionals could conduct similar experiments to assess the likely effects of their planned lighting design before installation on-site. Together with the perceptions of accessibility, sociability, hospitality, spaciousness, publicness, and social behavioural intentions, they can also investigate other impressions and behavioural choices sought through lighting design. The use of simulated images of real spaces, using real lighting fixtures, and simulating real luminous effects is an easier, and relatively more rapid and economic tool to test a lighting approach compared to mock-up and scale models. The results obtained are also considered reliable, as the literature review found that responses to digital images generalize well to those taken on-site in the actual environment (Stamps 1990). The responses to simulated luminous environments may also be compared to responses to the lighting conditions achieved after installation to test to what degree the on-site impressions and behaviours are consistent with the predictions and behavioural intentions anticipated. This evidence-based design approach can provide lighting professionals with guidelines on ways to improve their design to create more lively, accessible, and social public spaces after dark.

Given the study's exploratory nature, the results related to the lighting conditions that affect the impression of a street indicate that, to make a street feel larger, more hospitable, more public, and more sociable, uniform lighting with a horizontal and vertical distribution has the greatest effect, while to make it feel more private, non-uniform lighting has the greatest effect. To make a street a more socially inviting place to sit and meet others, the results indicate that uniform lighting with a horizontal and vertical distribution has the greatest effect. Lighting may be used to accent vertical and horizontal cues as well, such that people perceive that a space is more hospitable and sociable, which may increase its social use. The findings here suggest that uniform horizontal, vertical, and accent lighting can make a public street a more appealing place to socialize, and hence, encourage more social experiences in public spaces after dark.

References

Barazawa N, Hanyu K (2013) Effects of nonuniform lighting on the evaluation of spaces: a comparison between the situations with and without conversation. Jpn Psychol Res 55(3):273–283. https://doi.org/10.1111/jpr.12011

Baron R, Rea M, Daniels S (1992) Effects of indoor lighting (illuminance and spectral distribution) on the performance of cognitive tasks and interpersonal behaviors: the potential mediating role of positive affect. Motivation and Emotion 16(1):1–33. https://doi.org/10.1007/BF00996485

Boyce P, Eklund N, Hamilton B, Bruno L (2000) Perceptions of safety at night in different lighting conditions. Light Res Technol 32(2):79–91. https://doi.org/10.1177/096032710003200205

Butler D, Biner P (1987) Preferred lighting levels. Environ Behav 19(6):695–721. https://doi.org/10.1177/0013916587196003

Durak A, Camgöz Olguntürk N, Yener C, Güvenç D, Gürçınar Y (2007) Impact of lighting arrangements and illuminances on different impressions of a room. Build Environ 42(10):3476–3482. https://doi.org/10.1016/j.buildenv.2006.10.048

Fisher B, Nasar J (1992) Fear of crime in relation to three exterior site features. Environ Behav 24(1):35–65. https://doi.org/10.1177/0013916592241002

Flynn J (1988) Lighting-design decisions as interventions in human visual space. In: Nasar J (ed) Environmental aesthetics: theory, research and applications. Cambridge, New York, pp 156–169

Hargroves R (2001) Lighting for pleasantness outdoors CIBSE national conference 2001, pp 136–141

Hawkes R, Loe D, Rowlands E (1979) A note towards the understanding of lighting quality. J Illum Eng Soc 8(2):111–120. https://doi.org/10.1080/00994480.1979.10748578

Heijboer M (2013) seeing things in a positive light. Investigating the effects of color temperature of light on sociality and de-escalation. Master of Science in Human Technology Interaction. Department of Industrial Engineering & Innovation Sciences –Eindhoven University of Technology.

Hendrick C, Martyniuk O, Spencer T, Flynn J (1977) Procedures for investigating the effect of Light on impression. Environ Behav 9(4):491–510. https://doi.org/10.1177/001391657794003

Herzog T, Flynn-Smith J (2001) Preference and perceived danger as a function of the perceived curvature, length, and width of urban alleys. Environ Behav 34:653–666. https://doi.org/10.1177/001391602237250

Küller R, Ballal S, Laike T, Mikellides B, Tonello G (2006) The impact of light and colour on psychological mood: a cross-cultural study of indoor work environments. Ergonomics 49(14):1496–1507. https://doi.org/10.1080/00140130600858142

Miwa Y, Hanyu K (2006) The effects of interior design on communication and impressions of a counselor in a counseling room. Environ Behav 38(4):484–502. https://doi.org/10.1177/0013916505280084

Nasar J, Bokharaei S (2016) Impressions of lighting in public squares after dark. Environ Behav 49(3):227–254. https://doi.org/10.1177/0013916515626546

Nasar J, Jones K (1997) Landscapes of fear and stress. Environ Behav 29(3):291–323. https://doi.org/10.1177/001391659702900301

Newsham G, Richardson C, Blanchet C, Veitch J (2005) Lighting quality research using rendered images of offices. Light Res Technol 37(2):93–112. https://doi.org/10.1191/1365782805li132oa

Stamps A (1990) Use of photographs to simulate environments: a meta-analysis. Percept Mot Skills 71(3):907–913. https://doi.org/10.2466/PMS.71.7.907-913

Veitch J, Newsham G (1998) Lighting quality and energy-efficiency effects on task performance, mood, health, satisfaction, and comfort. J Illum Eng Soc 27(1):107–129. https://doi.org/10.1080/00994480.1998.10748216

Chapter 6
Interacting with the Social Human-Scale Lightscape

Abstract This chapter presents two case studies of experience prototyping that has been designed as experimental lighting probes. They were conducted both in the field and in the laboratory to explore and evaluate design ideas. The chapter has an exploratory nature, addressing a series of different perspectives in the realm of the psycho-social effects of lighting in the outdoor environment: the perception of intimacy and publicness along with the territoriality and personalization of moving people passing through a space as well as social proximity/distance and social behaviours/interactions. The qualitative analysis was based on observations documented with videos and photography, and audio-recorded, semi-structured interviews. Those were compared to obtain a deeper understanding of the influence of lighting on the social agency in the urban environment. The emerging preliminary insights of this research could provide practical design guidance for architects, landscape and lighting designers to develop computationally advanced lighting solutions that integrate the psycho-social effects of lighting in nocturnal lightscapes.

Keywords Experience prototyping · Social agency · Human-scale lighting · Interactive lighting

6.1 Introduction

The lighting design with a human and social intent aims at creating new opportunities for social encounters and engagement of people after nightfall. The lighting systems should enable the accessible use of the space, improve acquaintance and affiliation, support different uses through different lighting scenarios that promote social activities, and enhance the urban experience. The LED lighting technology has been digitally connected through sensors that enable lighting monitoring and management. Nowadays, the new smart-lighting systems consist not only of lighting but also are equipped and integrated into an electronic infrastructure, including sensing devices for detecting people's presence, quantity, movement, proximity, and posture, as well as many other details. This is opening the opportunity to design luminous atmospheres that are adaptable and flexible in relation to different users, social uses and atmospheric impressions that a place could communicate: lighting can be either

D. Casciani, *The Human and Social Dimension of Urban Lightscapes*,
PoliMI SpringerBriefs, https://doi.org/10.1007/978-3-030-57165-8_6

functional or evocative, defining visual cues, and be synchronized with social activities. This could have an influence on humans' psycho-physiological well-being (Karlicek 2012). Since the lighting system would be embedded with data and information on the real lives of individuals by entering their social infrastructure, its design demands to take into account social, psychological, aesthetic and functional features. The lighting performances allowed by these technological advances need to be supported by complex design decisions that should take into account the appropriateness of lighting to the context, by considering the physical settings, the architectural volumes, and the functions of the space, but also the well-being of individuals. Prototyping the experience of interactive lighting for social agency is fundamental for studying from the outset the relationship between people and dynamic/interactive temporary lighting scenarios. The research would highlight how lighting technology impact the impressions, reactions, behaviours and social activities of uninformed users in outdoor situations where safety is not an issue, since it is guaranteed by the familiar surroundings and the presence of other people.

The aim of these experiments was primarily to design human-oriented lighting performances, through dynamic and interactive public lighting, based on the qualitative perception of the visual environment, making them adaptable to different social uses. More than this, the objective of the experiment was to understand which thoughts, sensations and impressions come to people's minds in relation to dynamic or interactive lighting in the urban environment. In particular, the agency of lighting that is able to transform the intimate and private impression to a public, formal and detached one has been investigated in outdoor settings. Implicit interactions with different lighting scenarios (Ju and Leifer 2008) have been tested to support different social activities and behaviours for sociality explorations (Casciani 2014).

The two studies investigated lighting conditions in relation to different social settings in two different specific applications and were conducted in the field and in the laboratory. The first was focused on the transit, and explored the perception of intimacy and publicness along with territoriality and personalization of moving people passing through a space; the second study, instead, focused on a static situation of the encounter of two people and explored intimacy and publicness as well as social proximity and behaviours, along with territoriality perception. The following were the research questions:

- Which lighting features could positively influence the perceived atmosphere of a social night space?
- How could artificial lighting define a higher trust relationship between people and space?
- Which reactions and behaviours do dynamic or interactive lighting define? At what extent is lighting impacting people's behaviours in their normal routine?

The scope is to answer a more general question with practical implications for architecture and design of social atmospheres inside public outdoor areas: How should we design a lighting environment that adapts its behaviour to support the current social setting or eventually change it in a more positive way?

6.2 Study 1—Methodology

The first study has been set up in a Living Light Lab at the Eindhoven Market Hall of the Metaforum inside the campus of the University of Technology (TU/e). The scope was to explore the lighting influence in space territorialisation and personalization (Fig. 6.1).

6.2.1 Lighting Stimuli: Design and Description

The lighting system available in the Living Lighting Lab of the Market Hall was used to design different dynamic and interactive lighting scenarios that could provide people with contextual information translated into lighting events. Lighting variables

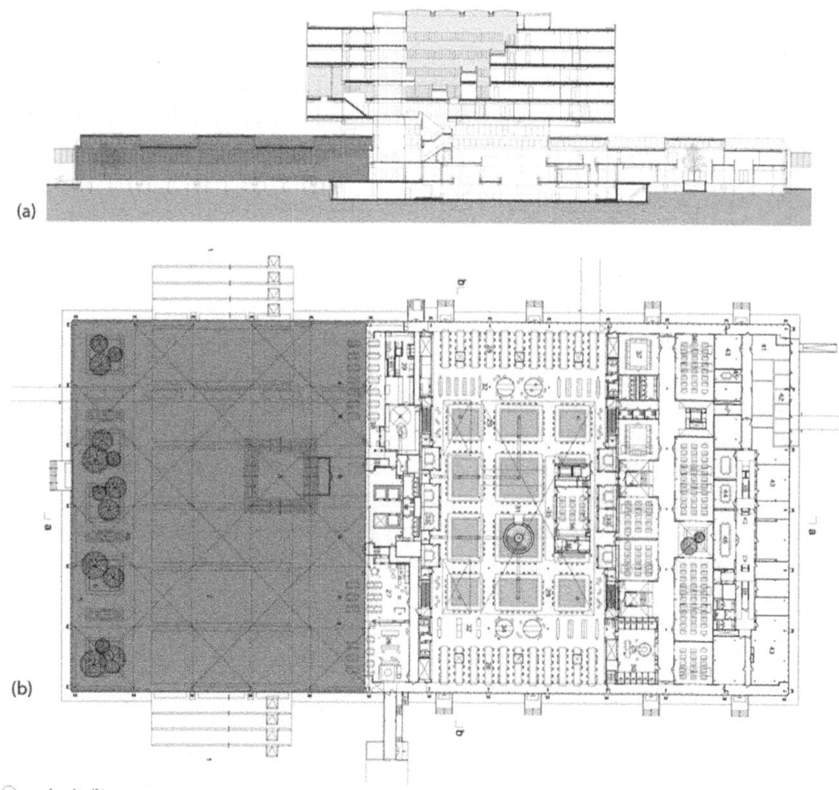

Fig. 6.1 Section AA (**a**) and plan (**b**) of the Metaforum with the Market Hall in dark grey. *Credits* Ector Hoogstad Architecten (2012)

Fig. 6.2 The architectural-oriented lighting scenario

were manipulated to define more or less noticeable lighting effects and three lighting scenarios were designed from more functional to more evocative ones:

- Ambient lighting scenario: static control lighting;
- Architectural-oriented lighting scenario: dynamic lighting on architectural details;
- Human-scale lighting scenario: interactive lighting echoing people's movements.

6.2.1.1 Ambient Lighting Scenario

Ambient lighting was set up when people were not physically present in the space, were not detected and the building was not in use. This lighting scenario was characterized by a static scene in warm white lighting and uniform lighting distribution, with decreased lighting levels. This lighting scene was static and was working as the control scene.

6.2.1.2 Architectural-Oriented Lighting Scenario

Animated spotlighting was designed and prototyped to focus dynamically on several architectural elements of the structure. The ambient lighting was left as default: warm, dim and uniform. The background vertical glass surface was lit with neutral CCT and high-intensity lighting and was fading with a very slow rhythm. The corridor of the space was lit with high-intensity and neutral CCT, highlighting the passage in fast succession. The pillars of the structure were lit with high-intensity spotlighting, performing slow dynamics of fading randomly in sequence. The illuminance contrast ratio between general lighting and dynamic lighting effects was designed for a strong and noticeable distinction (Fig. 6.2).

6.2.1.3 Human-Scale Lighting Scenario

Based on interactive lighting, people were used as unconscious dynamic switches and were enabled to illuminate the portion of the space they were using. The overall luminous performance was shaped by the social use of the space and, as a consequence, the resulting illuminated space perfectly reflected the way the Market Hall was used. Slow and subtle lighting dynamics that orient and follow the way people moved in the space were designed with a simple causal interaction. Lighting was

Fig. 6.3 The human-scale lighting scenario

echoing the way people used the space in response to different actions. The lighting interaction was generally structured on pre-programmed cycles of call-and-response between humans and lighting, considering the presence of one single individual, two individuals or a group of individuals using the space. Each person was detected, and they triggered a lighting reaction in relation to his/her position in the space and his/her behaviours. Lighting was made welcoming by increasing the warm light levels upon people detection, and served as a guide by orienting and showing the path. Lighting was also programmed to create a festive light event when two familiar strangers met in the space (Fig. 6.3).

The space was equipped with two sensorized modules that were placed where the main transitions and passages of users were observed. Each sensorized module was equipped with a ultrasonic proximity sensor ("Ultrasonic Sensor Product Catalog—MaxBotix Inc." 2019), a wireless Xbee antenna, a programmable open-source hardware electronics platform board, Lithne (Magielse and Offermans 2013), and a rechargeable battery pack for self-autonomous positioning in the space. The small-beam of the radar was quite precise in locating people and monitoring a small portion of space. Besides, the computational system behind the hardware was designed to auto-regulate and to face both anomalous situations and multiple users of the space by balancing the lighting performance or prolonging its duration.

6.2.2 Participants and Procedure

The experimental session lasted about 15 days. Every scenario was tested during five weekdays, for approximately one hour per day (7:00–8:00 p.m.). Over the duration of the experiment, more than 50 people per day were observed. Eighty interviews were conducted, with passers-by experiencing the dynamic and the interactive lighting tests. Participants were gathered in the space and therefore were traditional users during their routine activities. The age range varied between 21 and 50 years old, with a relatively evenly gender distribution.

A qualitative assessment approach was employed to collect data based on interviews and observations. The observation of activities and behaviours of the participants were based on ethnographic studies and fieldwork techniques used for understanding public spaces (Goffman 1959; Hall 1990; Whyte and Whyte 1984). These were focused on body language, gestures, head movements and walking trajectories. The researcher, as an observer, was located in a hidden but prominent position that allowed for the monitoring and video recording of the space used by people. The observation was followed by semi-structured interviews to collect first-hand reports about people's impressions of the prototyped lighting experience. A random sample of the people passing through the space was selected for interviews (lasting about 20 min). People were approached with an initial question, asking if they had noticed the lighting in the space ("*Have you noticed the lighting in the space?*"), followed by a questionnaire about the lighting experience in relation to personalization ("*Do you feel that the lighting is behaving according to what you were doing?*") and expectations ("*Is it something new for you or you would expected it?*"). If people responded they had not noticed the lighting performance, they were asked to experience it in a passive way, by watching the lighting scenario with other people in the space. Interviews were audio recorded to be transcribed and to review the content lately.

6.2.3 Analysis Methodology

A systematic approach for evaluating people's behaviours and reactions toward lighting was elaborated using a summarizing chart. In particular, in this behavioural checklist, five levels of interaction were devised: passing by, viewing and soft reacting, subtly interacting, directly interacting, multiple interactions and passive interaction. Observations were documented with videos and photography. Audio-recorded, semi-structured interviews were transcribed in quotations according to their representative and descriptive qualities. The quotations were arranged in clusters for the analysis. A reflexive approach was used to further develop the qualitative analysis of the interviews (Alvesson and Sköldberg 2009).

6.2.4 Discussion

The information extracted from people's interviews was examined within the observations to identify key themes and patterns of behaviours and people's impressions in relation to dynamic and interactive lighting scenarios. The collected data were clustered and discussed in terms of level of consciousness, expectations, territorialisation and personalization.

6.2.4.1 Level of Consciousness

Half of the passers-by were noticing the dynamic lighting scenario. The transformation of their behaviour was visibly observable because they turned their gaze and their heads toward the lighting system (up above) several times and reduced their walking speed. If alone, people directed their gaze toward the lighting system; if in groups, people were discussing the lighting system and testing if there was a form of interaction between them and lighting. The other half that didn't notice any lighting performance had their gaze directed toward the pavement the majority of the time (Davoudian and Raynham 2012) or toward their mobile phones. Eighty percent of the interviewed participants reported to have noticed the dynamic lighting related to the architectural elements of the Market Hall. The luminous dynamic environment determined different degrees of interest. Peripheral awareness occurred when people were not noticing lighting in a conscious way but just recognizing that "*lighting is sometimes moving*" or "*is not always on*". Curiosity occurred when people interviewed were asking for more information about the lighting events. Differently, focused awareness appeared when people demonstrated a conscious perception of lighting transformations that were considered "*pretty interesting*". In all the cases, people used the space in the way they were used to, without any change in their behaviour due to the dynamic lighting system.

The interactive lighting was highly noticed. As derived from the interviews and observations, people were aware of the interaction with lighting with different levels of consciousness. Peripheral awareness occurred with a small group of people that ignored the presence of lighting even if it presented strong illuminance ratios and faster lighting movements. The explanations were that lighting was taken for granted "*…for me it's normal, I think this lighting is already here.[…] I think it has to be here*", but also the hurry and the focus on other activities blurred the focus on lighting: "*I'm really out of mind. If I do not focus I really do not recognize anything*", "*sorry my mind was out of thoughts*" and "*because we were talking, we didn't see anything*". In those cases, people were invited to watch the interaction to see what was happening with other people. They became interested in what they missed and expressed immediate and involuntary, positive surprise and interest in both their facial and verbal expressions "*now I noticed it, so cool!*" and "*wow, I should have noticed that. Nice*".

Subtle unconscious perception occurred when people didn't show any reaction toward lighting but, when asked, described to have noticed a subtle luminous change: "*I was really out of my mind. Something was blinking when I was walking through the space*". In these cases, the lighting interaction was considered positive because it was not disturbing but just accompanying the respondent's principal activity: "*It was different, not just like regular, something unusual. I didn't notice it but in my mind I saw it, very subtle. I think that is good that it didn't attract my attention*".

A conscious perception of lighting changes occurred, causing a subtle interaction with lighting through body engagement and a slowing in pace. Respondents were often not able to describe the lighting changes very clearly, but used terms like "*sort of dynamic*", "*intensity of the light fluctuating a little*" or "*changes the intensity of the*

Fig. 6.4 Subtle unconscious and conscious perception of interactive lighting. Images taken through an hidden video-camera

light". A group of people was also interpreting the lighting changes as a personalized performance: "*I felt that light went on when I was walking through the space*", "*I noticed that the light became stronger when I walked, when I came out of it. It's like it follows me*", "*I thought that the light were switching on. Did they follow me? I had this perception*" and "*I noticed that it flash when we were going out it seems like someone that was taking a picture of us*". (Fig. 6.4).

Direct interaction occurred with a minority of participants, who changed their behaviours and started to talk about lighting. Once the link between people and lighting was recognized, they were testing if the lighting was following them through the space. When asked, they reported that having lighting follow them was positive both for aesthetical and environmental reasons.

6.2.4.2 Expectation

The level of acquaintance with dynamic and interactive lighting scenarios was heterogeneous: lighting was considered unexpected but not unconventional. The fact that lighting was activated by people was recognized as technologically possible and desirable since it was already experienced in interiors. Differently, participants stated to have experienced interactive lighting in outdoor common public spaces on very few occasions.

6.2.4.3 Territorialisation and Personalization

The influence of lighting was investigated in terms of space personalization and people territorialisation. The dynamic lighting scenario is composed of a series of timed lighting scenes that highlight elements of the architecture. Despite this, people have tried to find deeper evocative reasons for these dynamics. The most meaningful explanation was the personal influence on lighting, even though there was no linear, causal effect between presence or movement and lighting effects. This misunderstanding has indeed determined higher interest and pleasantness in respondents, while the light dynamics related only to the architectural elements have not been considered functional or worthy in the evaluation of the space.

The interactive lighting scenario was positively experienced as a presence detector, by following users' movements with slow and subtle lighting events occurring in a linear, causal way for helping to navigate the space. People were detected by two sensors, which in turn triggered the lighting in relation to their position and behaviours.

The majority of the interviewed participants reported a positive feeling about responsive lighting, often smiling when answering. Interviews highlighted different levels of neutral to positive impressions and approval: "*Whatever I don't care, I'm so tired I do not care*", "*Neutral*", "*Kind of positive*", "*Positive*", "*I like that*", "*I think it is nice*", "*Positive*", "*Positive and fun*" and "*Cool*". The fact that lighting was on when people went out of the building was appreciated the most when the lighting system was perceived in close proximity and by single individuals (more than people in groups). Participants related this positive feedback mainly to sustainability and evocative functionality of being recognized as users of the space: "*It is sustainable and you get a fresh feeling when you get out. Because you get outside of the building and your brain is very tired after the 9 h of study. Normally you are inside, studying by yourself, you are very much in the world within yourself. And then you get outside and you have a feeling something is interacting with you in this intermediate space, and then you get outside in the real world*".

A smaller amount of people noticed that lighting was also following the path during their walk in the space. When noticed and recognized as a direct casual effect of their presence and movement, these personal activations of lighting were experienced with positive feelings: people reported a sense of control and personalization: "*I feel I was the one doing that*", "*I thought it was that I did it*" and "*You think that something is changing because of you. You feel like powerful*". The reasons for the positive reactions to these lighting responses was very often related to having more light: expressions like "*Lighter*", "*More light*", "*increasing intensity*" and "*from dark to light*" were the most reported ones. The interactive lighting that was showing the path was associated to a positive feeling and no negative feedback was reported in relation to "being under the spotlight". In this regard, on a few different occasions, people tested if lighting was more personal by following people. This topic of the following spotlighting was quite recurrent and explained as a desire to have light on-demand on a personal basis against darkness.

Interactive lighting was found to be significant in personalizing the space and giving a sense of control, evocating a positive company meanwhile having a reassuring power. Hence lighting determined an impression of subtle management and active personalization of the luminous atmosphere, defining a subtle relationship with people through unconscious and not-invasive perception. Even implicit, the interaction with lighting was found to contribute to restoring an intimate connection with the space. Besides this, the direct bodily interaction with lighting increasing both levels and personal control, as found by Hans and de Kort (2012), ensures the security perception and comfort of the individuals, without creating embarrassment.

6.3 Study 2—Methodology

Inspired by the studies of Kobayashi (2013) and Magielse and Ross (2011), this prototyped experience was focused on understanding sociopetal/sociofugal behaviours, social proximity, social appraisal and lighting control consciousness occurring in relation to lighting. The experiment was conducted in a full-scale space, arranged in a laboratory room located in the Design Department of the Politecnico di Milano. The dimensions of the room were 4 m × 6 m. The walls and ceiling were white, and the 30 cm × 30 cm ceramic-covered floor tiles were medium grey (20% light reflectance). Environmental conditions such as air temperature, humidity, furniture and layout were kept almost constant. As evident from the plan (Fig. 6.5a) and the side-elevation layout (Fig. 6.5b) of the experimental room, the room was left intentionally unfurnished, with only the sheltered interactive lighting prototype, a table and two chairs for the participants.

6.3.1 Lighting Stimuli: Design and Description

Three lighting scenarios, different in terms of the tonality of white (3000 K-5000 K) and distribution (direct narrow spotlight—direct medium spotlight—direct/indirect), have been prototyped in a micro-environment. The lighting system is composed of a shelter with integrated lighting modules and sensors able to monitor presence and body posture of a couple of participants.

The lighting experience was designed to have three different interactive scenarios, which depended on the monitoring of people in the space of the shelter. If no one was intercepted by the sensors, the direct narrow spotlight in warm CCT (3000 K) would light the table with an horizontal illuminance E_h of 160 lx. If the bodies of the couple were sensed as leaning backwards (in a position of social detachment), the luminous atmosphere would change to cold white (5000 K) and direct/indirect

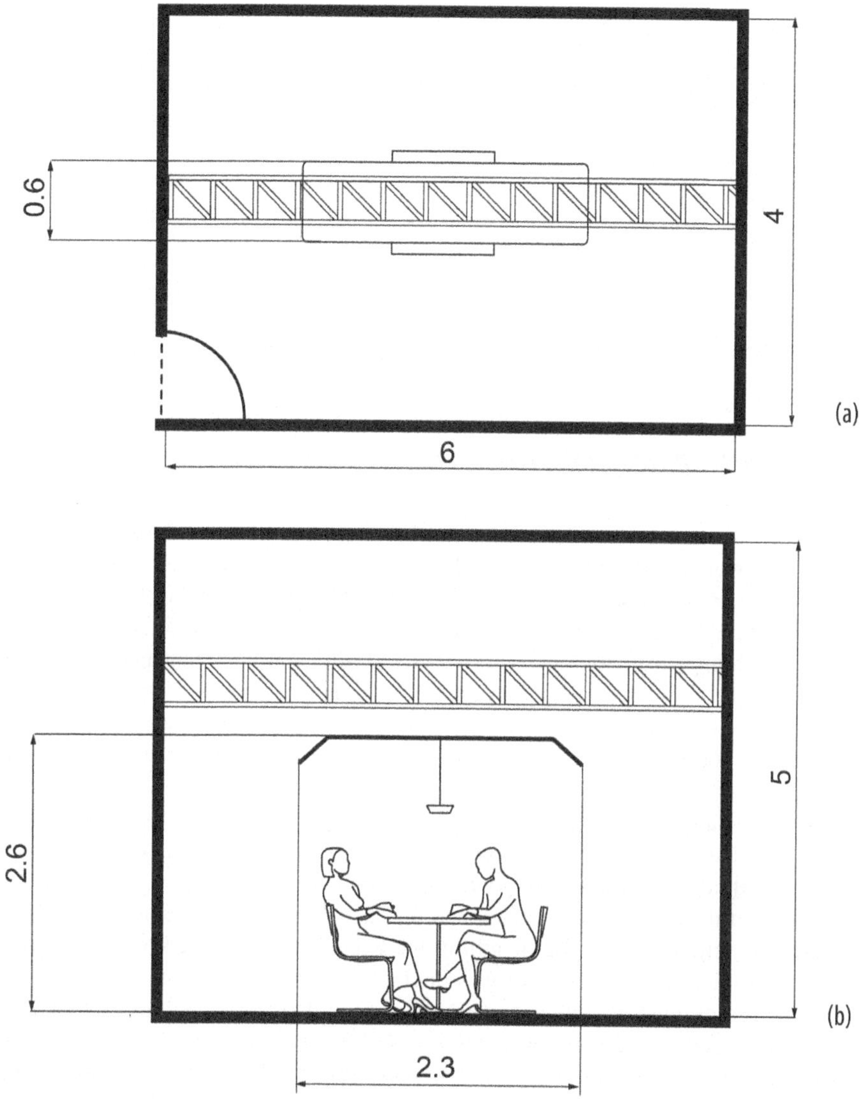

Fig. 6.5 Plan (**a**) and side elevation (**b**) of the experimental room

lighting distribution, reaching an E_h on the table of 1540 lx. If the bodies of the couple were detected as leaning forward (in a position of social proximity), the luminous atmosphere would change to a direct warm white medium spotlight with a CCT of 3000 K and an E_h on the table of 1070 lx (Fig. 6.6).

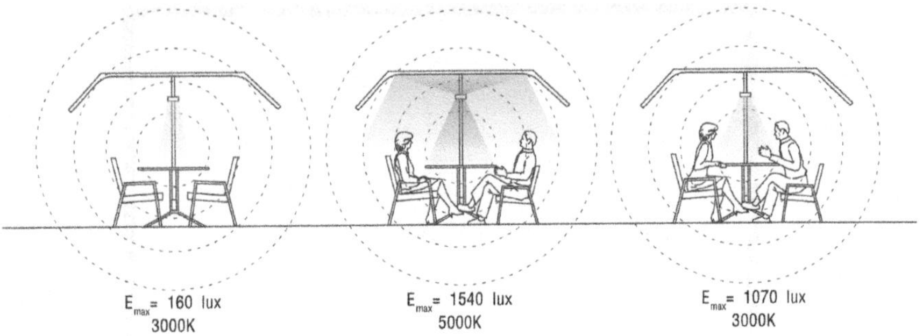

Fig. 6.6 The luminous experience set up to investigate sociopetal/sociofugal behaviours and social proximity

6.3.2 Participants and Procedure

The experimental test was applied to 20 (14 females and 6 males) voluntary participants, who were divided into 10 pairs. Eighty percent of participants were students, and the average age was 25 years old. The nationality breakdown of the group was 50% Italian, 15% Turkish, 5% Lebanese and 5% Russian. The recruited participants were invited, in pairs, to join the experiment and, after administering and completing the informed consent document, they answered a demographic questionnaire. Subsequently, they entered the experimental room and were asked to sit at the table, to feel at ease and to perform role play of different social activities in two different non-controversial situations. Each group was asked to enact the role play in two opposite, socially distancing (e.g., meeting a stranger in the city, discussing a job, being a student under examination) or approaching situations (e.g., known friends meeting and talking, counselling, journalist interview).

The order of the presentation changed for each experimental session to counterbalance the sequence bias. The length of the experiment depended on the duration of the two role-play sessions and was enough for lighting adaptation. During the role-play prototyping, participants were video recorded. At the end of each experimental session, the participants were interviewed about the noticeability and evidence of the lighting transformation and implicit interactions, the lighting evoking social approach or distance, and the lighting influence on interpersonal relationships.

6.3.3 Analysis Methodology

Role playing was a method that, through the use of a physical interactive lighting system/environment, allowed for the testing and capturing of the users' emotional experiences with different lighting scenarios and social situations. Feedback and insights about the experience were analysed, combining interviews and video recordings. A selection of interview quotations was chosen according to their representative and descriptive qualities. The quotations were arranged in a cluster as the basis for the analysis. To counterbalance and compare what was derived in the interviews with what actually happened during the experience, video recordings were analysed through the observation of behaviours.

6.3.4 Discussion

6.3.4.1 Evidence of Implicit Interactions with Lighting

Despite the lighting system not expressing evidently how it was controlled, the difference between the lighting scenes was recognized and sometimes considered too visible: "*Well you can notice pretty much and I suppose that it is slightly too much. It's a little bit disturbing. The changing between the two scenes should be smoother*". More subtleness and delicateness in the lighting interaction was desired so that the difference between the scenes goes more unnoticed.

A group of participants interacted explicitly to further explore the meaning of the lighting transformations. Participants were not constraining their behaviours, but rather showing interest and testing the lighting system through explorative behaviours and gestures, which made visibly evident their desire for personal control and occupation of territory. The majority of participants were guessing, during the experience, the reason for lighting changes. Their interpretations were various and ranged between the categories of connectedness and empathy: the interviewed participants believed that the interaction happened through voice-volume recognition: "*When I was lauder I noticed a flash*", "*I noticed that it is changing while we were laughing, maybe the tone of voice higher or lower*", or movement and personal posture monitoring: "*it changes in relation to the distance we have*", "*I thought it was related to the hands over the table or was about our movement but I couldn't really understand why!*", and "*Tone of voice and movement, when you were moving toward me I have noticed the lighting was changing. Instead I was static and I didn't change the lighting*". In addition to this, they also thought that the system was able to sense their emotions, thoughts and mood, not only by monitoring what they felt but also to anticipate their needs and express publicly their feelings: "*Is it detecting out the emotion or feelings? Is it able to detect the heat when you feel embarrassed?*", "*This is reading our thoughts!*", "*Lighting is changing when I don't like what is happening*" and "*I thought it was understanding the momentary mood*". A small group of participants thought that the lighting was controlled by another person (e.g., the researcher) through a remote control.

During the experience, different levels of interaction occurred between participants and lighting: indirect engagement with lighting, direct interaction between people who started to talk about the meanings of the light changes, and direct interaction between people and lighting triggered by the researcher during the interview phase.

6.3.4.2 The Lighting Influence on Social Proximity/Distance and Social Interaction

The majority of the participants found the interactive lighting system to be supportive during the role play and useful to assist in the performance of fictional social activities without being intrusive or disturbing their attention (Fig. 6.7). During the interviews, people addressed the warm lighting as more intimate, comfortable and cosier, enhancing "*communication and closeness in psychological terms*", defining a more intimate atmosphere, which fit more the social situations. The warm spotlighting conditions were found to shape a closer and more private atmosphere in intimate situations: "*…it is so romantic!*", "*when the light was lower and warmer I felt more comfortable. When the light was bright I felt more detached*", "*I think that the warm lighting is creating a more intimate relationship. It is like you are in a speed date bar, restaurant or a situation like this were the warm lighting is creating a closer relationship between the two people involved*" and "*The warm light feels like cosy.*

The feeling of being surrounded by darkness and being less exposed to external sight allowed respondents to talk more openly about personal information. This lighting condition suggested and supported more privacy, intimacy and closeness, and also delimited a personal territory: "*So you focus only on the person and you look at the eyes!*" When the dim and circumscribed lighted environment was brightened, it invited less-intimate interactions. In this regard, lighting was a material immaterial tool that was perceived as signalling the transition between one mood to another, as was also noted by Knapp et al. (2014). Cold direct/indirect lighting was found to be more formal and detached, shaping more distance between individuals by showing the faces and the surroundings; people felt more exposed in this luminous condition, which lit an open, shared territory: "*The cold lighting is helping in maintaining the distance; warm lighting is helping in creating the atmosphere but it depends on the context*", "*when you have the cold light, you see the face of the other, but in the same time you are more exposed and you are in a less intimate lighting condition that makes you talk less in the private affairs and subjects. The other warm light is more private*" and "*It's like to be too much revealed, exposed to have the cold lighting around that everybody can see you in an intimate situation*". The lighting features influence the social proximity and social interaction and are perceived as context. The social and behavioural processes are perceived as linked not only to the lighting features but also to the physical setting (environment) and the socio-spatial context (de Kort 2015).

According to the studies of Kobayashi (2013), gender seems to differentiate impressions, attitudes and behaviours about social lighting in closeness situations. One of the male-female couples among the participants affirmed to have felt embarrassment during the experiment, in particular, thinking that lighting was displaying and sharing publicly their personal and social information. They were also disturbed by the intimate lighting condition, which was considered not appropriate for the social activities they were playing. The belief of public disclosure of feelings disturbed the participants: "*I think it is kind of weird because the system it is showing the feelings*". Conversely, with male-male and female-female couples, lighting was not creating

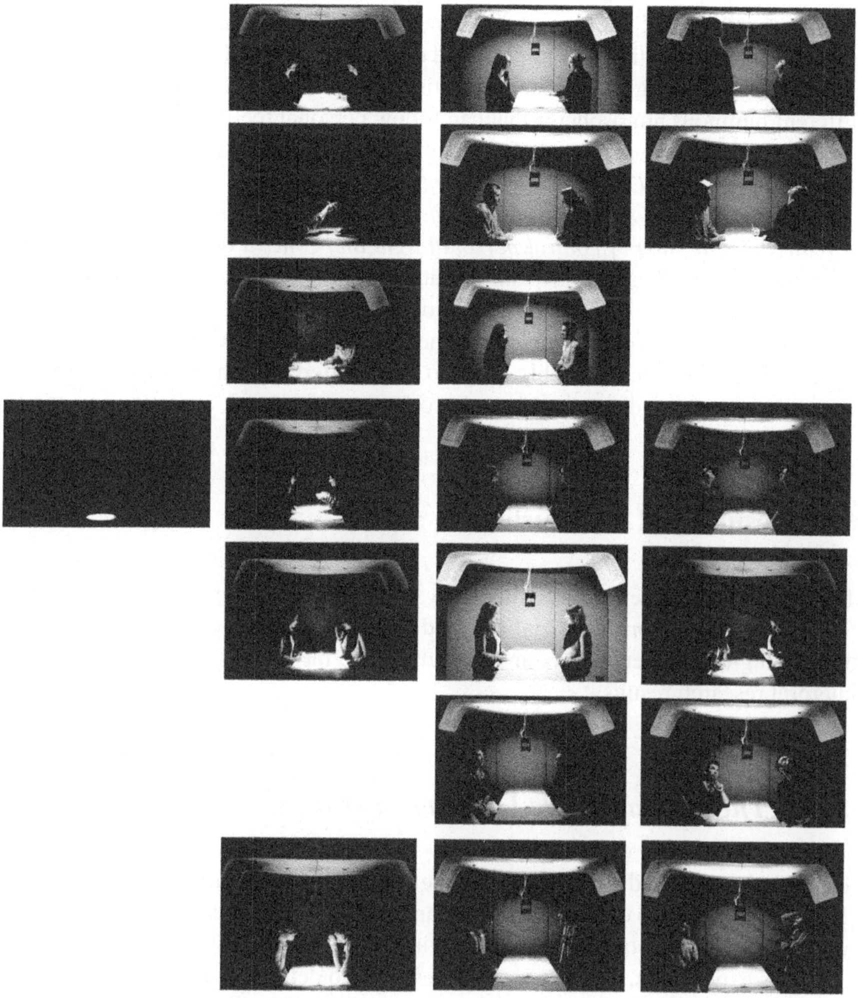

Fig. 6.7 Screenshot of the role-play videotapes: the first column shows the scene with no people; the second column shows several moments of social closeness; the third column shows situations of social detachment; the fourth column shows people exploring the lighting systems and its functions

these kinds of disturbances. Personal and shared behaviours that were disclosed by lighting transformations were not causing evident discomfort also because the sensors used were not considered intrusive like cameras.

Interviews revealed that the majority of the participants enjoyed the lighting system, which was found as effectively accommodating the luminous atmosphere in relation to the proxemics impressions of people. The social agency of lighting in this experiment was determined mainly by cultural association that had been accumulated through past experiences. In fact, many times in the interview, people mentioned "the lighting recalls about" previous lighting atmospheres, and this association was often used to decide if lighting was appreciated or not in the played social role.

The idea of having an interactive lighting system that is controlled to determine different luminous atmospheres that can follow personal feelings and emotions was openly and positively approved of, depending on the context of application: "*I think that could be a good idea to have a control of the lighting so to make the space comfortable or more scary depending on the atmosphere you want to create*" and "*I really like the idea of control personal light and zone in order to accommodate personal lighting*". In addition to this, participants described also the applicative difficulties of this interactive lighting scenario, in particular, considering what happens in situations with groups of people and when people show opposite social needs. The system should be able to flexibly adapt to these different attitudes and decide to be supportive for social distance or social proximity.

6.4 Results and Conclusions

The conclusion gathered throughout the previously described experiment can be formulated as a theory of social-oriented lighting both in terms of findings from the lighting-design perspective and in terms of the methodology used to collect and interpret research results.

6.4.1 The Human-Scale Lighting Interaction

Through the case studies explorations, the results showed lighting not to be the sole determining factor to define a more human- and social-oriented space. Lighting can act as a feature that complements the environment to its social quality and use.

In this experiment, lighting was not designed to determine particular behaviours different from the natural ones or to interfere with the normal use of the space, but instead, was designed to support it in a functional, social and evocative way through the environmental atmosphere perception. The influence of certain '*lightscapes*' (Bille and Sørensen 2007) on people's behaviours should be always read as mediated by contextual, cultural, environmental, personal and social factors. Despite this, the results showed that certain luminous atmospheres and lighting variables could have a social evocativeness and could contribute to setting a more human-scale lightscape in terms of safety, intimacy and hospitality in outdoor settings. Through the different experimental prototypes, an understanding of people's appraisals, preferences and impressions of the luminous atmosphere of outdoor spaces was elaborated.

In the first study, people were found not to be passively reacting to external lighting conditions, even though they were goal-directed: they were consciously and unconsciously interested in lighting and subtly transformed by it not in terms of behaviours but in the way the outdoor environment was perceived and described.

This transactional subtle mechanism of people and lighting was not resulting in a new set of behaviours, such as people going out more at night, meeting socially in the space and so on, but was aimed more at re-establishing people's psychologies, expectations and values, and creating new cultures of public space.

The observations indicated that interactive public lighting was perceived as more engaging in comparison to dynamic architectural lighting, which was found meaningful only if correlated to people activities. These findings represent important contributions in the exploration of the social and aesthetical dimensions of contemporary lighting in the urban environment.

People expressed a desire for a closer relationship with lighting in terms of physical proximity, bodily interactions and personal responsive effects. The experience of personalization of the atmosphere of the outdoor space, defined by the "*It is because of me*" expression, was the main reason for technological positive acceptance. Some preliminary insights about people's relations to interactive lighting performances focus on the on-time feedback, on the recognizable lighting response, on the simple-to-understand lighting interaction without cluttered information and on the subtle lighting effects in terms of rhythm.

According to lighting standards such as the EN 13201 (2016) and CIE 115:2010, uniformity was probably one of the most important factors to consider/guarantee when designing urban lighting. In the case of dynamic or interactive lighting performances, the uniformity factor is not guaranteed, and the most important lighting variable is the control of illuminance contrast ratios between lighting provided over people, when they are intercepted, and the surrounding ambient lighting. In the experiment, ambient lighting was always set up as more than 0 lx (the minimum was 6 lx at the ground level) and the illuminance contrast with the lighting effects was manipulated between 2:1 and 10:1, to observe the level of consciousness and disturbance on people. In this respect, people never felt discomfort because of glare and over-illumination.

For many of the participants, interactive lighting was an unexpected and novel occurrence for an outdoor environment that could have a wider application, not only from a functional point of view (increasing energy efficiency of the system), but also from an evocative point of view (having lighting that follows). In this respect, bodily interaction with lighting that increases the level of personal contact, control and transformation of the city luminous atmosphere was welcomed. As previously found by Hans and de Kort (2012), providing noticeable lighting levels in the near proximity of the user and adapting it in real time by following people's direction and movements was found to guarantee safety and comfort with a pleasant feeling of surprise and excitement.

Lighting responsiveness to people's activities defined a more human-oriented urban environment. The interactive lighting that followed people's movements and activities in the space with a subtle but noticeable illuminance ratio was perceived as a human-oriented lighting feature for a positive, social, ethical, evocative and meaningful experience. Lighting was found to turn on not only a relational interaction with the space but also a responsible behaviour, providing more lighting just when the space was used. In addition to this, lighting was found to possibly turn on special gatherings and micro social interactions between familiar strangers.

The second study demonstrated that lighting has the agency of manipulating people, with a subtle influence on social behaviours, inducing background reactive and proactive human–light interactions (Ju and Leifer 2008). Social behaviours are not only socially based or bound up on cultural association but are also rooted in luminous atmospheres, conceived as the intermediate state between light, the environment and human perception. In particular, warm white lighting and the lighting distribution in the space could affect the personal and interpersonal space requirements, along with the territorial and social behaviours. In this, past experiences, cultural sensitivities and individual taste have a determinant role in defining the social agency of lighting atmospheres. The results of the prototyped experience highlighted that the effect of lighting is delicate, especially when social activities take place. Even though the explorations demonstrated that lighting has a subtle influence on the social behaviours of the participants, it is still recognised in its rooted social meanings. Therefore, it is also unconsciously influencing and leveraging deeper social meanings. In this regard, light acts supportively of social behaviours in specific real or fictional social applications to accommodate or compensate for more private/intimate or public/detached situations. Light acts to enforce interpersonal relationships, and to support social negotiations, and contributes in communicating proxemics information and defining more socially including or excluding environments. The public visual disclosure of personal and social information should be also considered when designing socially adaptable lighting scenarios and system due to the possible negative drawbacks in terms of experience.

In terms of methodology, to design socially adaptable public outdoor lighting with people in mind, starting from an ethnographic and social exploration of the space through an experience prototyping case studies could be useful both for lighting design practice and research aims (Buchenau and Suri 2000). The approach of developing experiential scripts for digital urban lighting solutions was found to be an interesting and challenging approach to inform future interactive lighting installations, which should be positively perceived by people.

The multiple approaches of data collection using observation techniques, interviews, audio and video recordings were also very useful for the comparison of data, to reinforce or reject the insights from different points of view. The experiments afforded the opportunity to employ some of the research and evaluation techniques developed by the social sciences. In particular, qualitative methods such as ethnographic observations and interviews of people utilizing the space in a natural setting with natural behaviours were quite uncommon and unique in the field of lighting design but are quite usual in the context of environmental psychology and design. The insights gathered by this qualitative approach were used to discuss aspects of people's interpretation about impressions and appraisal and also to reflect on people's reactions and behaviours toward lighting. On the other hand, the research is highly context dependent, and the experience and impressions of lighting and urban space are highly situational, depending on the context where the studies were conducted. The findings and insights are not generalizable and valid for all urban spaces. Despite this, the case studies present an initial contribution to the design of socially adaptive public lighting in contemporary cities, which advocate for a deep investigation of different environments and various other situations. Experience like this seems to be crucial in the future development of the so-called smart cities where lighting scenarios and human behaviours should be designed and tested in advance in order to influence, positively, the social use of the city. In fact, the research about the influence of lighting on sociality can concur to create better and more meaningful experiences through the use of new technologies. The use of the interactive lighting in micro-environmental prototypes confronted people with a new level of awareness about the future possibilities of lighting and choices that were not present before. The emerging results of this research could provide practical design guidance for architects and landscape and lighting designers to develop computationally advanced lighting solutions that integrate the psycho-social effects of lighting in the outdoors.

References

Alvesson M, Sköldberg K (2009) Reflexive methodology: new vistas for qualitative research, 2nd edn. Sage, London

Bille M, Sørensen T (2007) An anthropology of luminosity. J Mater Culture 12(3):263–284. https://doi.org/10.1177/1359183507081894

Buchenau M, Suri J (2000) Experience prototyping. In: DIS '00 Proceedings of the 3rd conference on Designing interactive systems: processes, practices, methods, and techniques, pp 424—433. New York City

Casciani D (2014) Esplorando la dimensione sociale dell'illuminazione urbana. Luce 308(52):93–97

CIE 115—2010 Lighting of roads for motor and pedestrian traffic.

Davoudian N, Raynham P (2012) What do pedestrians look at night? Light Res Technol 44(4):438–448. https://doi.org/10.1177/1477153512437157

de Kort Y (2015) Light on and in context. TechnischeUniversiteit Eindhoven, Eindhoven

Ector Hoogstad Architecten (2012) MetaForum TU/e. Nieuwe universiteitsbibliotheek en faculteit Wiskunde and Informatica in en op historische machinehal. Rotterdam

Goffman E (1959) The presentation of self in everyday life. Anchor Books, New York

Haans A, de Kort Y (2012) Light distribution in dynamic street lighting: two experimental studies on its effects on perceived safety, prospect, concealment, and escape. J Environ Psychol 32(4):342–352. https://doi.org/10.1016/j.jenvp.2012.05.006

Hall E (1990) The silent language. Anchor Books, New York, N.Y.

Ju W, Leifer L (2008) The design of implicit interactions: making interactive systems less obnoxious. Des Issues 24(3):72–84. https://doi.org/10.1162/desi.2008.24.3.72

Karlicek R (2012) Smart lighting—more than illumination. In: Asia communications and photonics conference (ACP). IEEE, Guangzhou, China

Knapp M, Hall J, Horgan T (2014) Nonverbal communication in human interaction. Wadsworth/Cengage Learning, Belmont

Kobayashi S (2013) Lighting effects on unconscious human behaviors. In: 10th international symposium on sustainable healthy buildings annual spring conference of Korea Society of lighting and visual Environment. Seoul

Magielse R, Offermans S (2013) Lithne—a platform for interaction designers to develop interactive networked environments. In: 9th international conference on intelligent environments. IEEE, Athens, Greece

Magielse R, Ross P (2011) A design approach to socially adaptive lighting environment. In: Proceedings of the 9th ACM SIGCHI Italian chapter international conference on computer-human interaction: facing complexity, pp 171–176. Alghero
Ultrasonic Sensor Product Catalog—MaxBotix Inc. (2019) Retrieved from https://www.maxbotix.com/Ultrasonic_Sensors.htm
UNI EN 13201–2: 2016—Road lighting—Part 2: Performance requirements
Whyte W, Whyte K (1984) Learning from the field. Sage, Beverly Hills

Chapter 7
From Design Research to Design Practice of Urban Social Lighting

Abstract This chapter presents the results of a research and practice-based/educational experience focused on the development of social-oriented lighting solutions in a specific urban context of application: Piazza Leonardo da Vinci, Milano (Città Studi). This experience included research focused on surveying and reading the context, identifying user activities and pattern of use, experimentation, concept design and an iteration of development, simulation and testing toward the definition of future lighting scenarios and visions for the cultural, social, environmental and technological enhancement of the area. The chapter presents the methods and contents of the practice-based/educational experience, analysing the proposed lighting solutions in terms of human and social needs and the related lighting techniques and variables (CCT, coloured lighting, lighting fixture dimensions and design, and lighting distribution). The derived insights focus on the parameters used to create luminous ambiances for pedestrians but also on presenting the holistic approach toward social-oriented lighting design: the study of context and users enable the gaining of diverse and specific lighting solutions that boost the human and social experience. This approach also transforms the design process toward integration and collaboration with different disciplines and stakeholders.

Keywords Lighting design practice · Lighting parameters · Social lighting design · Social analysis

7.1 Introduction

7.1.1 The Urban Social Lighting in the Design Process

Social oriented lighting design is an emerging approach diffused with research and design projects aimed at experimenting more intelligent, people-centric ways of using urban lighting (see Chap. 1). Some seminal case studies of social oriented lighting projects have demonstrated to contribute in the positive perception of the city and its social use, generating and assessing meaningful urban experiences also in the practice of lighting design (Davoudian 2019).

D. Casciani, *The Human and Social Dimension of Urban Lightscapes*,
PoliMI SpringerBriefs, https://doi.org/10.1007/978-3-030-57165-8_7

The human and social-oriented lighting approach was found powerful in supporting social life in the night of the cities in both functional and evocative ways. This approach tailors lighting design to human and contextual needs, to support the diverse space identities and citizens' day to day life, to design inclusive urban spaces that increase the opportunities for citizens to use the public space at night and, as a consequence, to socially interact.

The review of the lighting practice reveal a trend of lighting designers, with their identity and individual practice and methodologies, that are shifting to the human and social urban lighting focus, without abandoning the traditional scopes of urban lighting. Their design purposes can be summarized in the following ones:

- functional lighting: assuring lighting compliant with normative;
- monumental and architectural lighting: enhancing architectural elements;
- scenography lighting: creating atmospheres to shape the nigh-time image of the city completely different from the daylight image;
- social lighting: focusing on people comfort and engagement;

The common aspect is the need of creating urban spaces for people, encouraging activity and psychological wellbeing after dark. This intent is achieved by reasoning in terms of quality, local context (meanings and information given by the space such as morphology and materials, shapes and geometry) but also people experience, perception and attention.

According to the studies of Linnaea Tillett (Maile Petty 2007), the majority of the lighting projects are not conceived to make a place more social or sociable through lighting as the determining factor, but lighting is certainly one of the different design opportunities to complement the urban outdoor environment to its social quality.

> Despite from the fact that lighting is for amenity, safety, accessibility, security and way-finding, an extra layer it can create is called experience. Extra light layer is used to make the experience. In our project we have tried to create a systematic understanding on how to make a space an experience where people want to be, want to go. This is related to social issues, to how to create a space where people want to go, that people are talking about and maybe that they are directly creating because you are allowing people to take control with the system. How to evaluate this experience layer is actually very difficult task for lighting designer. It is important because we should be able to state more precise environmental objectives than "create a mood" or "create an attractive space." (Keith Bradshaw transcript excerpt of a Talk given at the Light Symposium Wismar, 2012, speech "Light - Architecture - Space")

Indeed few lighting designer have described their lighting design practice into a formal knowledge, contributing to the creation of basic principles different from the ones of lighting engineering, to depict a cannon in the urban social lighting. With the scope of starting a connection between research, education and practice in the lighting design domain, this chapter present a case study where research and analysis methodologies were applied in the process of lighting design by a group of students. Along with the context analysis, a strong emphasis was put on the user investigation and in the social and human experience assessment and design. Social oriented lighting design criteria were used to support the process of design of more human-oriented experiences of the night-time urban environment. Based on the literature research (see Chap. 2), the human and social urban criteria were presented as a tool and differentiated in the following macro-categories:

- Functionality (visibility, visual comfort, visual acuity, glare, facial recognition, urban legibility, spatial recognition, visual orientation, way-finding, saliency and visual hierarchy of perception);
- Wellbeing (Security, safety perception);
- Ambiance and urban beautification (architectural, material and landscape enhancement, socially inclusive/inviting, relaxing, interesting and expressive, arousing and lively luminous atmospheres);
- Relationship/experience (personalization, socialization, interaction);
- Environmental issues (skyglow, light trespass, light profligacy).

7.2 Methodology

We present the results of a research and practice-based, educational experience focused on the development of social-oriented lighting solutions in the urban scale, taking into consideration either the social-oriented lighting criteria and tackling issues regarding the exploration of context and end-users to inform the design process in the very early stage. The preliminary research part was conducted under the framework of the Vivipolimi Research at the Design Department of the Politecnico di Milano with the particular focus of surveying and reading the context, identifying user activities and pattern of use, of Piazza Leonardo da Vinci (Città Studi) along with the Rector's Building of the Politecnico di Milano. The design part was conducted during a project work called "*POLIMI and Light 4.0: The future of Urban Lighting*". Focusing on cities and artificial lighting, students were asked to envision the nocturnal city of tomorrow and its relationship to both light and darkness, by defining future lighting scenarios and visions for the cultural, social, environmental and technological enhancement of the area. Including research, experimentation and concept development, this experience was aimed at enabling students to develop social oriented lighting solutions by testing design tools. This contribution presents the contents and methods of this research and teaching experience. The student learning results are reported highlighting the associative recurrences and relevancies between the proposed lighting projects, the selected social oriented lighting criteria and the lighting variables.

7.2.1 An Overview of the Research Strategy

The preliminary research phase has been set up to map the luminous situation of the area from the user perspective and to gather feedback from a contextual and user-oriented approach to define some insights and requirements for the preliminary lighting design development. It has covered the following aspects:

- Historical and architectural analysis, focusing on the style of the facades, through the analysis of volumes, voids and built parts, the study of ornaments of historical or decorative importance (protrusions, windows, overhangs, frames, etc.);
- Lighting analysis, surveying the position, orientation, typology, illuminance and luminance level and lighting distribution of the current lighting system;
- Perceptual analysis, capturing the view of pedestrian from different points of view from the various entrances and passages mainly pedestrian. Exploring the comparison of the day and night situation (existing project) through photographs to better understand the state of the art of the luminous atmosphere;
- Material analysis, measuring and assessing the relationship of materials with light (light reflectance and chromaticity).

7.2.2 An Overview of the Design Challenge

The Lighting Design for Outdoors project work was delivered to 30 students of the Master in Lighting Design and LED Technology at the Politecnico di Milano. The course was offered to both first level degree students (73%) and second level graduated students (27%). International students (10 males and 20 females) presented a heterogeneous background: (33%) architecture, (27%) interior design, (20%) product design, (7%) design and visual communication, (10%) fine arts, and (3%) electrical and electronics engineering. The course was conducted by the author in April 2020, over a period of one month and a total of 40 h delivered in a digital class. The methodological approach of the course was based on an introductory theoretical lecture (expository method) followed by the challenge-based practical part (discovery methods through hands-on experiences with a learning-by-doing approach). This problem/project-based learning approach confronted the students with a challenge to solve in the field of urban lighting design through a concept and design development that should provide an "*intellectual and technological*" solution for the nowadays complex urban illumination discussions focused on the human-scale lighting dimensions. Reflecting upon how societal attitude, technological opportunities and environmental challenges might change and shape the Urban NightScape required them to gather information from various resources and to analyse the context from different perspectives. The course has been divided in the following different main phases:

- an introductory theoretical lecture providing students with an accurate knowledge of urban lighting design principles to be applied in their design to help them in problem-solving;
- background analysis and synthesis research and presentation in terms of history, architecture and social use of the area;
- concept and ideation phase through a series of assignments to produce lighting sketches and photo-renderings, moodboards about inspirational lighting project, and the urban lighting concept masterplan;

- project layout development of the main features of the lighting design plan with a technical description/definition/design of the main features and lighting fixtures, through calculation and verification of the project;
- project presentation through the urban masterplan, sections and details with lighting fixtures legends and also through renders/representation of the lighting project, showing the difference between daytime, current night-time and night-time with the new lighting design.

Since experiential learning is a student-centred rather than teacher-centred approach (Briers 2005), the instructors' role was simply to facilitate a smooth process, monitor partial results, observe and stimulate observation/reflection, guide and help in the correction of misconceptions or misinformation. The output of the process did not focused on achieving well finished end-results. Instead, the emphasis was placed on stimulating the process that involves students in problematizing, reframing and iterating upon their design project (Cross 2010; Nelson and Stolterman 2003).

7.3 Results

7.3.1 Understanding the Social Layer

7.3.1.1 Daytime and Night-Time Pattern of Pedestrian Use and Spatial Functions

Piazza Leonardo da Vinci and Politecnico di Milano were set up in 1927. The square, which is the hearth of the neighbourhood Città Studi, extends over two hundred meters in length. It was completely redeveloped in 2015 to become a pedestrian area: the parking lots from the front of the Main entrance of the Politecnico were eliminated, new precious materials were located in the pedestrian paths and streets, the large ellipsoidal flowerbed in the centre was transformed into a large park area, the green areas have been increased by many square meters, with curated gardens and benches.

The exploration of urban daytime and night-time pattern of use and functions was aimed at depicting the human-oriented identity of the district (Fig. 7.1). It was implemented through a brief observational study, recording the use of the space at daytime and night-time Piazza Leonardo da Vinci, today, consists of:

- the historical part which is the brain of the area and hosts the University and the most ancient buildings of the Politecnico di Milano;
- the middle area with the green park and evergreen trees, a calming situation within the city's chaotic vibe;
- the surrounding residential area with private buildings.

The Piazza is eminently used by pedestrian (students, visitors, Politecnico's workers, inhabitants of the surroundings) as a walkable area: park paths, green areas and urban utilities such as the tram and metro station exits, but also the main square

Fig. 7.1 Urban daytime (left) and night-time (right) image of the Piazza Leonardo da Vinci

in front of the Rector's building that is an important aggregation spot, without forgetting the presence of cars in the surrounding spaces and so the safety/security issues in the pedestrian crossing areas. From the diagrams (Fig. 7.2) it is evident that the Piazza has a double rhythm: by day, it shows a pulsating vitality of people at work and study: they flow from the Metro and tram station (north side) to the Politecnico in the morning. Students use the green areas to eat and relax during lunchtime and afternoon. By evening everyone flows away and the area becomes less populated at night. Despite this, the square is perceived as a place of union and sharing for spontaneous aggregation of young people and students. It presents different levels of socialization after dark: most of the people gather around the central stairs' steps in front of the main building or in the benches to chat and enjoy the space. People use the central areas also to dynamically interact by playing sports such as jogging, skateboarding and playing Frisbee. During the winter evenings, the square is in use by the late workers and students but at night is poorly populated. Differently, the summer nights are populated with organized events to gather and enjoy the summer convivial public spaces.

7.3.2 Supporting the Envisioning Process of Lighting Design

7.3.2.1 From Analysing the Space to Synthesizing Requirements

The scope of design is to develop a strong, clear, communicative vision, that can be translated in concepts or strategies for the nocturnal urban space and then produced in a lighting project. The capability of transferring the historical, architectural, contextual and social analysis in a synthetic and incisive way was supported through the use of some analytical tools. Students were guided to articulate the analysis considering strengths (S) and weaknesses (W) as internal factors within the area and opportunities (O) and threats (T) stemming from external forces. The SWOT analysis was conceived as a strategic planning technique to enables to structure the decisions in the initial concept design stage when planning the project. Mainly referred to the development of business plan, the SWOT technique was intended as a starting point

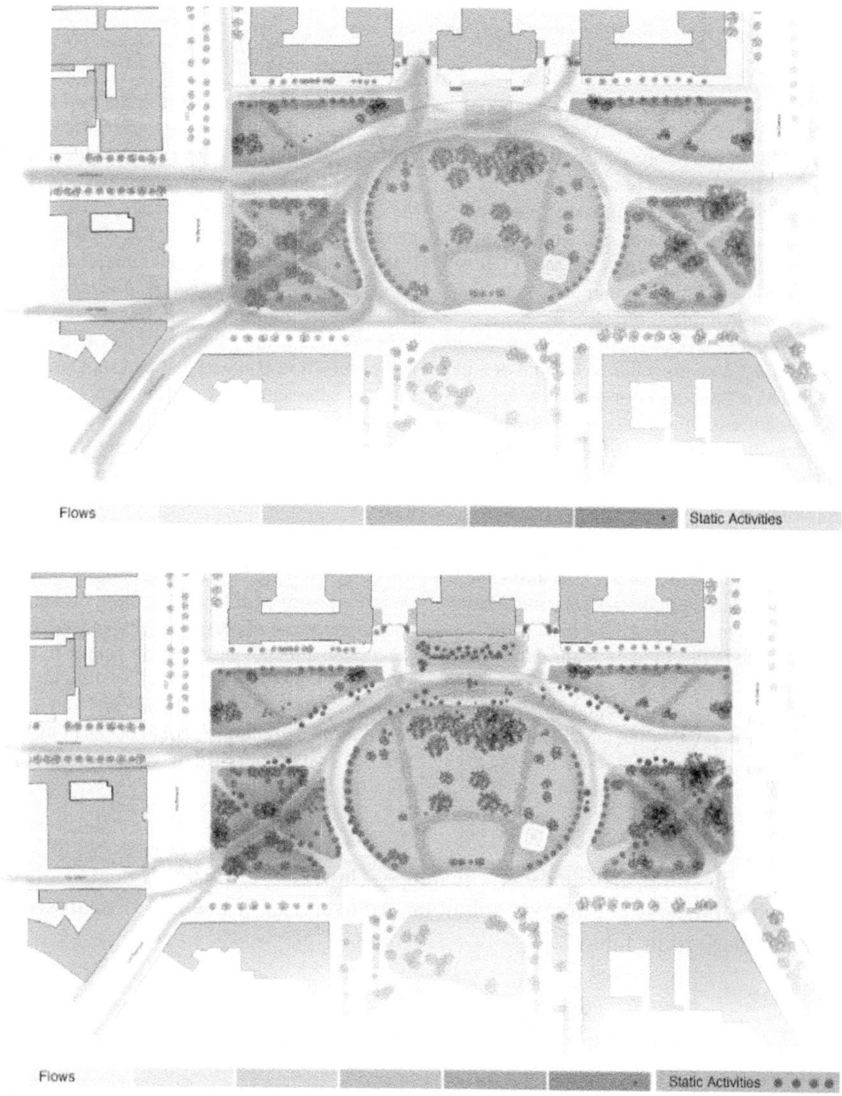

Fig. 7.2 Urban daytime (up) and night-time (down) comparison of the pattern of use and functions

for the students to synthesize the analysis in the preliminary stage of the process of decision-making.

The single responses of the groups of students were clustered in topics and reformulated in macro-categories, by rephrasing and synthesizing contents and reporting some quotations. As a general overview, students referred to the importance of illuminating human and social experiences, connecting the old University Campus with the city, supporting pedestrian activities and special events, and highlighting the

Table 7.1 Aggregated results of the SWOT analysis

Strenghts	Weaknesses
"*Strategic hub*" in the University campus showing good connectivity, transit and permeability from public transportation and thus pedestrian use due to large pedestrian and cycle paths, central gathering square and an exceptional big green area and landscaping which act as a "*barrier from the roads*". It is also an icon scenic location either since "*the main rector building is the most iconic construction for Polimi*", either because of its "*fashionable Neo-classical style*" that define the area as a "landmark for the entire campus". These features define a very popular pedestrian area that allows large movement of people either in the daytime and in the night hours	The area seems to be underlit in the evening either in the green areas, in the pedestrian paths and accesses and in the lateral buildings. These volumes of darkness do not adequately mark the spaces to satisfy its functional ("*visual perception of the surroundings*"), wellbeing ("safety perception") and social features ("*aggregation, engagement and encounters*"). In general the lighting situation is not flexible to be adapted to different functions, activities and events. It is also very flat and standard, thus not enhancing the experience, nor giving emotional and attractive expressiveness or sense of appropriation and character to the area and the Neoclassical rector's building
Opportunities	**Threats**
City—Campus reconnection: the large flow of people and the vitality of the area allow the space to have great potential via a purposeful lighting design to increase the night-time socialization in this outdoor semi-public area. Good "*dedicated, suggestive and hierarchic lighting*" could make this square a "*lively and recognizable*" point of the city, boosting "*creativity and appropriation*", "*improved social security, better navigation*"	Lack of proper lighting distribution and zone left completely in the darkness in the pedestrian paths increase the feeling of danger and decrease the sense of security and safety perception and, not encouraging people in the space by night along with creating difficult navigation of the space. "*Vandalism*" and "*long-term degradation*" could occur in the space within the absence of carefully and tailored good lighting

typical architectural and landscape features. They also critically reflected on the limitations of the current lighting solutions. An overview of the SWOT analysis has been summarized in Table 7.1.

7.3.2.2 From Framing Ideas to Visualizing Concepts

Designers should communicate through representation tools aimed at envisioning and presenting, to the eyes of different stakeholders, the final achievable results of the proposed lighting project. Students were asked to document and visualize their concept design using photographs, sketches, images and plans, to refine and create a clear visualization for their proposals.

They had to visually and communicatively emphasize the different layers of lighting to evidence the functional aspects, the architectural opportunities, the social elements through the lighting composition and quality, the juxtaposition of colours and CCT (correlated colour temperature), the brightness hierarchy, and the integration of lighting controls.

Proper working tools are necessary to help communicate, to get everyone aware of how to implement and to understand the lighting design concept. To guide the student to structure and give a shape to the lighting concept for the entire area of interest, they were required to sketch and render (either by hand or digitally) perspective images of main nodes, paths and landmark of the project and also to work on the Lighting Concept Master Plan (Fig. 7.3). This tool is meant to give guidance to the development of the lighting plan and to have a general overview of the complete area, comparing the different zones, evidencing, through representation, the lighting levels and distribution, CCT, coloured lighting and lighting effects, explained through a legend and specific notes.

7.3.2.3 From Identifying Meaning to Assessing Impacts

A specific "*Urban Social Lighting Design Radar*" tool was implemented for analysing the lighting design performances before and throughout the process of design. This tool has been conceived to assess the current luminous situation but also to support the improvement of the approaches toward a social-oriented night-time lighting design.

Data could be visualized as plotted in a web that allowed the designers to evaluate subjective (evaluative) as well as objective (measurable) parameters in an analytical way, to help inform the project and to compare the beforehand and afterword situation.

From a designer/professional point of view, this tool could be used to assess a specific situation and compare it with the possible impact the future projects could have, particularly enhancing the social and environmental aspects. In assisting the evaluation of the as-is situation, the tool is important, together with the SWOT analysis, to derive insights related specifically to the social-oriented features of the current lighting situation. In proposing a new lighting project by setting goals and requirements, the tool has been also hypothesized to be supportive to inform the conceptualization phase.

In this specific challenge-based project, we took the single answers from all the different work of students and considered the responses (average). The resulting aggregated comparison between the as-is situation and the new lighting design intents were plotted in the radar tool and described through statistical analysis.

The current lighting situation was considered fair in terms of functionality, not particularly good or bad in terms of visibility (mean: 3.3; st.d.: 0.8), legibility and making sense of the space (mean: 3.2; st.d: 0.9), spatial affordance (mean: 3.2; st.d: 0.9), way-finding and orientation (mean: 3.1; st.d: 1.05), facial recognition (mean: 3.1; st.d: 0.7) glare avoidance (mean: 3.2; st.d: 0.8), and visual acuity (mean: 3.1; st.d: 0.9). Accessibility was evaluated as almost good (mean: 3.6; st.d: 1), meanwhile saliency and visual hierarchy of perception, meaning the recognition of functions and main important elements in the space were evaluated as almost poor (mean: 2.3; st.d: 1.4). The environmental aspects were considered almost fair, neither good neither bad in terms of light trespass avoidance (mean: 3.2; st.d: 1.3), light profligacy avoidance (mean: 3.2; st.d: 0.9) and the skyglow reduction (mean: 3.1; st.d: 0.7). Similarly, in

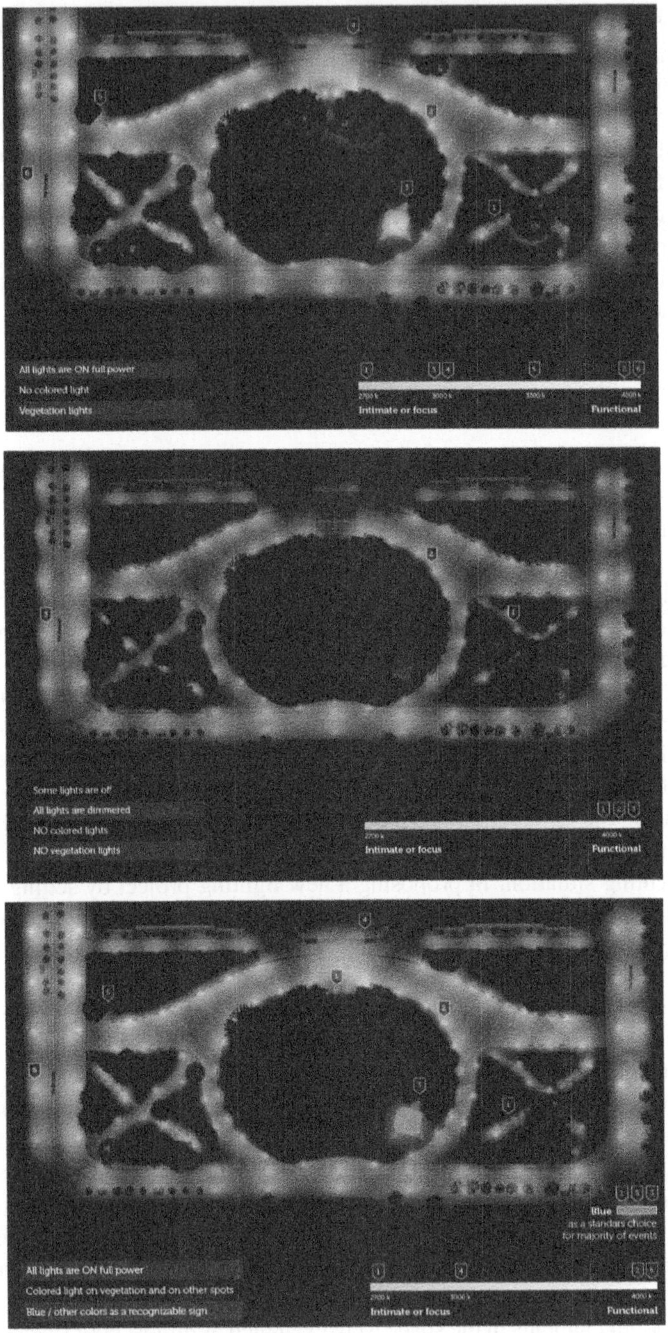

Fig. 7.3 Scenes of the lighting masterplan of the project "Some light remains". Credits to Sofia Mari, Francesco Dal Molin, Gherardo R. Rossi

terms of wellbeing, the space was evaluated almost fair in terms of sense of security (mean: 2.8; st.d: 1) and safety perception (mean: 2.7; st.d.: 0.6). In relation to the ambiance and urban beautification, the evaluation scored low values, lacking the fundamental qualities for representing a social-oriented space at night, in particular, in terms of material (mean: 2.5; st.d.: 1.2), architectural and landscape enhancement (mean: 2.3; st.d.: 1.2), and in the definition of socially inclusive/inviting (mean: 2.3; st.d.: 1.4), relaxing (mean: 2; st.d.: 1.1), interesting and expressive (mean: 2.1; st.d.: 1.4), arousing and lively atmosphere (mean: 2.1; st.d.: 1.4). The relationship and experience aspects gathered at night in the space were considered poor in terms of entertainment (mean: 2.1; st.d.: 0.9), socialization (mean: 2.1; st.d.: 1.2), and personalization (mean: 1.6; st.d.: 1.3).

The proposed solutions, in response, aimed to increase all the aspects related to functionality, environment and wellbeing, but above all the qualities and criteria related to ambiance, urban beautification, experience and relationships. In terms of functionality, saliency and visual hierarchy of perception (mean: 4.6; st.d.: 0.5) along with way-finding and orientation (mean: 4.5; st.d.: 0.5) were proposed as very indispensable criteria to be achieved with the new lighting project. Facial Recognition (mean: 4.3; st.d.: 0.5), visibility (mean: 4.2; st.d.: 0.6), spatial affordance (mean: 4.1; st.d.: 0.3) and legibility (mean: 4.1; st.d.: 1.1) scored high values and were considered necessary for the new project to be achieved. Accessibility (mean: 4.3; st.d.: 0.7), visual acuity (mean: 3.6; st.d.: 1.1) and glare avoidance (mean: 3.6; st.d.: 0.7) were considered important but not necessary to be enhanced, also because the current situation was considered already almost good. In respect to the environmental aspects, light profligacy avoidance was considered extremely indispensable (mean: 4.6; st.d.: 0.5), followed by light trespass avoidance (mean: 4; st.d.: 0.7). Skyglow reduction did not changed so much from the as is situation (mean: 3.4; st.d.: 1).

In terms of ambiance and urban beautification, architectural and landscape enhancement (mean: 4.6; st.d.: 0.5) were considered to be extremely indispensable more than material enhancement (mean: 3.6; st.d.: 0.5). Between the different achievable atmospheres, the socially inclusive/inviting one scored the highest values (mean: 4.8; st.d.: 0.3), followed by the interesting and expressive (mean: 4.3; st.d.: 1.1), relaxing (mean: 4; st.d.: 1.1) and arousing and lively atmospheres (mean: 3.8; st.d.: 1.2). Lighting was considered necessary in order to increase the socialization (mean: 4.1; st.d.: 0.6) and the personalization (mean: 3.8; st.d.: 1). Lighting for achieving entertainment (mean: 3.2; st.d.: 1.3) was considered important but not necessary. Finally, the sense of security (mean: 4.4; st.d.: 1) and safety perception (mean: 4.4; st.d.: 1) were considered necessary in the perception of the space after dark (Fig. 7.4).

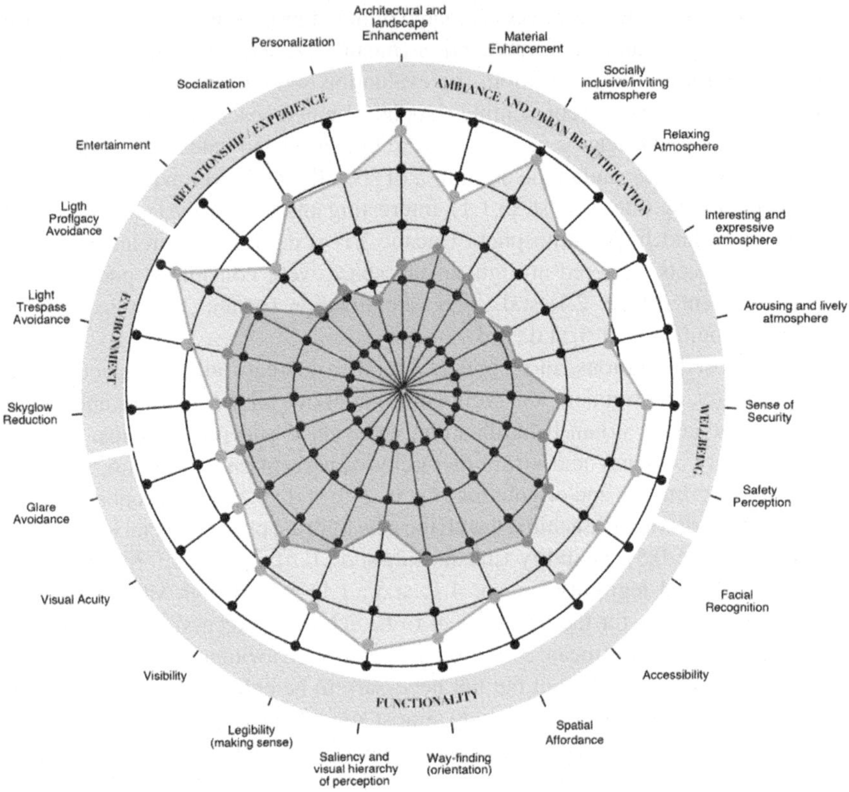

Fig. 7.4 Human and social lighting radar comparison between the evaluated as is situation and the proposed solutions

7.3.3 *Informing the Human and Social Lighting Design Features*

The proposed solutions were analysed to gather insights and trend in the selection and use of lighting features to fulfil human and social-oriented criteria that were considered as important requirement of the project.

Students were asked to list/declare the three most important human and social lighting criteria/needs and then to describe the three most important lighting techniques they would use to accomplish those needs. Answers were clustered in the most recurrent criteria: socialization and socially inviting/including atmosphere, safety and security, way-finding and orientation, space/function legibility, and architectural enhancement were mentioned by all the groups.

In relation to the definition of a socially enhanced space, different lighting parameters were mentioned to create a sense of intimacy, socialization and interaction: increasing the lighting levels "*on more dynamic areas of social interest*", using "*lower levels*" and "*warm CCT to create a sense of intimacy*", creating lighting scenarios that "*attract the attention*" toward the landscape and also for special events by using colours, specific layout of lighting fixtures that could be tilted to create "*tunnel of lights*" or task-lighting "*to highlight the benches*".

Security and safety perception were achieved through good level of "*luminance on the paths*" and also "*providing lighting in the green areas*". Left in darkness in the current situation, highlighting the "*hierarchy of the streets*" with different lighting fixtures' heights and luminous levels. Parameters such as "*uniformity*" and "*neutral CCT*" were correlated to the "*functional layout of lighting fixtures*".

In terms of wayfinding and orientation, many proposals were focused on achieving a perceptual perspective toward the rectorate building, creating a visual hierarchy between light and dark and guiding the eyes. This could be achieved through "*lighting levels adaptation*" toward the main focal point, a "*forced perspective*" using poles of different heights, layering lighting for a "*visual hierarchy*" among different areas, use "*linear light effects on the ground to enhance perspective*".

Space/function legibility was achieved through the "*differentiation of CCT*" and "*pattern of light*" enabling also the "*branding of the space*". In relation to this issue, architectural and material beautification were achieved through the use of "*correct CCT for each material*", lighting levels and beam aperture to "*enhance the floor pattern*", trees and to highlight "*the hierarchy of the architectural details*". Coloured lighting would be used "*according to the events*".

To understand how those social oriented lighting criteria were achieved during the development of the project, the proposals were analysed by detailing the areas of pedestrian interests (1—Small paths, 2—Benches, 3—Square in front of the Rector's building, 4—Pedestrian streets, 5—Pedestrian and car street, 6—Green areas) (Fig. 7.5) in relation to the used lighting variables (CCT, coloured lighting,

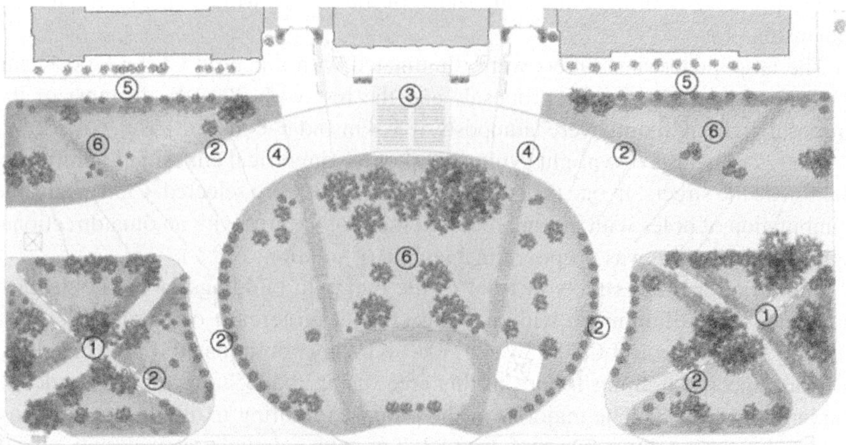

Fig. 7.5 Areas of interests: 1—small paths, 2—benches, 3—square in front of the Rector's building, 4—pedestrian streets, 5—pedestrian and car street, 6—green areas

height of the lighting fixture and lighting distribution). Results were analysed aggregated to find descriptive trends in the practice of lighting design to achieve more sociable, safe and secure, legible luminous atmospheres, helping navigation and enhancing the human/social-scale Table 7.2.

In the small path, the majority of the proposals used 3000 K (55%), followed by 4000 K (33%) and one solution used RBG lighting. Most lighting fixtures were mounted at low heights: 33% of recessed luminaires installed at the ground level, 44% bollards (0.35, 0.9 and 1 m) and 23% poles of 4 and 6 m. Besides, some poles were customized to be tilted toward the passers-by or to create a luminous tunnel effect. Considering the light distribution, 66% were downlight with an asymmetric (elliptical) and symmetric beam angle, the 34% were floor washer with a medium and wide beam aperture.

Lighting scheme for the benches was specified only by 44% of the proposals. Among these, half used warm CCT (2900 K and 3000 K) and half used 4000 K. In terms of height, half of the solutions were installed at 0.5 m (located under the benches) and the other half were pole solutions of 2 and 2.3 m.

Conceived as the main focal point of the area, the square in front of the Rector's building was illuminated at 4000 K (57%) and with warmer CCT (2900 K and 3000 K) for the restant 43%. In some proposals, the two whites were used together and the warmer CCT was selected for ground and stairs' recessed lighting fixtures, where social encounters occur. This combination of two different lighting languages was also evident in the selection of heights: poles, with heights comprised between 6 and 9 m and downlight asymmetric lighting distribution were adopted in all the proposals. In 5 proposals, lampposts were combined with ground level lighting fixtures, such as recessed floor washers with narrow beam or recessed linear luminaires with an omnidirectional lighting distribution. The scope was, on one side, to increase human proxemics and sense of intimacy through small size pools of light and on the other side to properly illuminate the wide shape of the square without losing the human dimension.

The large pedestrian streets were illuminated with a neutral CCT (3700 K and 4000 K) in the 66% of the proposals and the rest with 3000 K. In most of the proposal, lighting fixture were lampposts of 3–4 m and 5–6 m (one case), 6 m (36%), 7–8 m (27%) with a downlight symmetrical or asymmetrical emission to uniformly illuminate the streets. In one proposal, bollards of 1 m were selected. Conversely, the combination of poles with ground recessed lighting fixtures with an omnidirectional lighting distribution was proposed in a couple of solutions (18%).

Pedestrian and car streets in front of the two twin buildings of the Politecnico di Milano were illuminated with a slightly higher preference of warm CCT (55% at 3000 K) than neutral CCT (33% at 3000 K). Other solutions did not considered to place specific lighting fixture in this area, using the lighting of the facades of the buildings (22%). The majority of the proposed lighting fixtures were lamppost with an height ranging between 4, 6 and 7–8 m, with a downlight asymmetrical and symmetrical distribution. In two proposals ground recessed luminaire were selected with prevalently floor washer with narrow beam.

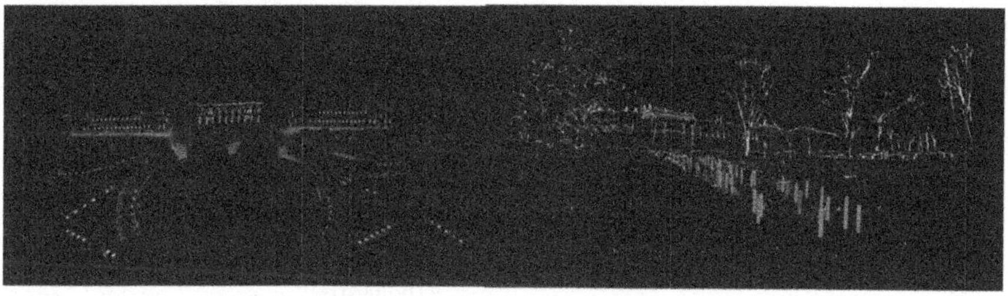

Fig. 7.6 Visualization through photosketching of the lighting proposal. Credits to Maria Elena Marcon, Isabella Cassisa, Carlotta Renzi

To increase security and safety perception, along with defining an attractive landscape, green areas were illuminated using both white (warm and neutral CCT) but also multi-coloured RGB or RGBW lighting (50%). Colours were meant to be used for special events. Most of the lighting fixtures were located as ground recessed (0 m) and were used to up-light the trees (mainly evergreen) with different beam apertures depending on the shape and the desired effects. Besides, 4 m tall poles with a tree-shaped morphology were selected, along with customized lighting systems and 0.25 m tall downlight were installed to accent the small bushes (Fig. 7.6).

All proposals were elaborated with lighting scenes adapted to the timing and activities/events, by dimming 70% after midnight or turning off some portion of the lighting scheme. In one case, movement and proximity sensors were used to create responsive lighting scenes to the users.

7.4 Discussion and Conclusions

This research presented the results achieved during a preliminary research followed by a challenge-based workshop in form of a learning-by-doing educative experience. The experience included research, experimentation, concept design and an iteration of development, simulation and testing. The course focused specifically on allowing beneficial and reciprocal exchanges between tutors and participants, understanding and handling the complexity of designing lighting for outdoors through the use of several tools to bring out lighting concepts and projects. A portfolio of 10 lighting design proposals were developed.

The analysis of the proposed lighting solutions explored mainly the pedestrian experiences and focused on how human and social needs could be addressed by lighting design techniques and variables. This knowledge is embedded in the proposed solutions as implicit and instinctive practice of novice lighting designers. The goal of the study was to investigate what light characteristics designers use to create certain atmospheres, and how these light characteristics are related to a certain atmosphere. A similar approach was used in a study conducted by Seuntiens and Vogels (2008). Here, solutions were not driven by professional lighting

Table 7.2 Aggregated results of the proposed solutions

Area	Group	CCT	Colour	Height	Light distribution
1—Small path	G1	3000 K	–	1 m (bollard)	Downlight—Asymmetric
	G2	–	RGB	0 m (recessed)	Floor washer—Medium beam
	G3	3000 K	–	0.9 m (bollard)	Downlight—Asymmetric
	G4	3000 K	RGBW	4 m (pole)	Downlight—Asymmetric
	G5	4000 K	–	0 m (recessed)	Floor washer—Wide beam
	G6	3000 K	–	6 m (pole)	Downlight—Symmetric
	G8	4000 K	–	0.35 m (bollard)	Downlight—Asymmetric
	G9	4000 K	RGBW	0 m (recessed)	Floor washer—Medium beam
	G10	3000 K	–	1 m (bollard)	Downlight—Asymmetric
2—Benches	G1	2900 K	–	0.5 m	Downlight—Asymmetric
	G2	4000 K	–	2 m	Omnidirectional
	G3	–	–	–	–
	G4	–	–	–	–
	G5	–	–	–	–
	G6	3000 K	RGBW	2.30 m (tilted)	Omnidirectional
	G8	–	–	–	–
	G9	4000 K	RGBW	0.5 m	Omnidirectional
	G10	–		–	–
3—Square in front of the Rector's building	G1	3000 K	–	7–8 m	Downlight—Symmetric
	G2	4000 K	–	8–9 m (tilted)	Downlight—Asymmetric
	G3	4000 K	–	6 m	Downlight—Asymmetric
	G4	4000 k	–	8 m	Downlight—Asymmetric
	G5	2900 K (stairs) 4000 K (poles)	–	0 m (recessed) 8 m (poles)	Omnidirectional Downlight—Asymmetric
	G6	3000 K (stairs) 4000 K (poles)	–	0 m (recessed) 6 m (poles)	Floor washer—Narrow beam Downlight—Asymmetric
	G8	3000 K (accent) 3000 K (poles)	–	0 m (accent) 6 m (poles)	Floor washer—Narrow beam Downlight—Asymmetric

(continued)

Table 7.2 (continued)

Area	Group	CCT	Colour	Height	Light distribution
	G9	3000 K (recessed) 4000 K (poles)	–	0 m (recessed) 9 m (poles)	Omnidirectional Downlight—Asymmetric
	G10	4000 K (recessed) 4000 K (poles)	–	0 m (recessed) 7/8 m (poles)	Omnidirectional Downlight—Asymmetric
4—Pedestrian Street	G1	3000 K	–	3–4 m and 5–6 m	Downlight—Asymmetric
	G2	4000 K	–	8 m (tilted)	Downlight—Asymmetric
	G3	4000 K	–	6 m	Downlight—Asymmetric
	G4	4000 K	–	6 m	Downlight—Asymmetric
	G5	3700 K (recessed) 4000 K (poles)	–	0 m (recessed) 8 m (poles)	Omnidirectional Downlight—Asymmetric
	G6	4000 K	–	6 m	Downlight—Asymmetric
	G8	3000 K	–	6 m	Downlight—Asymmetric
5—Pedestrian and car street	G1	3000 K	–	4 m	Downlight—Symmetric
	G2	–	–	–	–
	G3	3000 K	–	6 m	Downlight—Asymmetric
	G4	3000 K	–	0 m (recessed)	Floor washer—Narrow beam
	G5	3000 K	–	0 m (recessed)	Floor washer—Narrow beam
	G6	3000 K	–	6 m	Downlight—Symmetric
	G8	4000 K	–	4 m	Downlight—Asymmetric
	G9	–	–		–
	G10	4000 K (recessed) 4000 K (poles)	–	0 m (recessed) 7–8 m (poles)	Omnidirectional Downlight—Asymmetric
6—Green areas	G1	3000 K	–	0 m (recessed)	Uplight—Medium Beam
	G2	–	RGB	0 m (recessed)	Uplight
	G3	3000 K	RGBW	0 m (recessed)	Uplight
	G4	4000 K	–	0 m (recessed)	Uplight

(continued)

Table 7.2 (continued)

Area	Group	CCT	Colour	Height	Light distribution
	G5	3800 K (green) 4000 K (trees)	–	0.25 m (green) 0 m (recessed)	Downlight Uplight—Flood Beam
	G6	3000 K (trees)	RGBW	1 m (recessed)	Uplight—Flood Beam
	G8	3000 K (trees)	–	0 m (recessed)	Uplight—Washer Beam
	G9	4000 K (trees)	RGBW	0 m (recessed)	Uplight—Medium Beam
	G10	3000 K (trees) 4000 K (general)	RGBW	0 m (recessed) 4 m (general)	Uplight—Various Beam Downlight—Symmetric

designers with a lot of experience in the field of lighting design for the outdoor. They were students of different design disciplines, previously educated in lighting design, guided to explore the context and the users, guided to criticize the actual lighting situation and to synthetize the project's aims. The results of this research showed that they were able to elaborate specific luminous ambiances based on social oriented lighting criteria, by means of different lighting characteristics.

It could be argued that this kind of practical knowledge is subjective, biased and not systematically achieved. Conversely, the used methodology allowed to define some descriptive trends and insights into the different proposed solutions, through an articulated analysis. The results were analysed to understand the relation between the social oriented lighting criteria and the lighting techniques, technologies and parameters (CCT, coloured lighting, lighting fixture dimensions and design, and lighting distribution) used in the projects. Furthermore, it can be expected that more combinations of lighting parameters exist that can result in the same or even more enhanced luminous ambiances for pedestrians. Nevertheless, the findings of this study could be a good starting point for further explorations about the relation between the different lighting features and the luminous ambiances for pedestrian in the practice of lighting design.

7.4.1 Luminous Ambiances for Pedestrian

An intimate and socially inclusive atmosphere was designed with lighting supporting the social context and relationships through a balanced and hierarchically controlled use of contrasts, warmer CCT and lower sized lighting fixtures. Particularly in small paths and around the benches, proposals focused on the selection of enveloping and human-sized lighting.

A reassuring atmosphere with lighting providing safety perception, improving urban legibility, accessibility and way finding was ensured in the other wider streets and in the square in front of the Rectorate Building. A neutral CCT and higher lampposts were selected to create a hierarchy, a focal point of view with higher brightness to orient and guide the eyes in reading the functions of the space. Neutral CCT was also selected to match the same CCT of the surrounding streets' luminaires. Besides this, lighting fixtures at human size were located also in those social areas to recreate an intimate situation with warm CCT.

A lively and memorable atmosphere with lighting supporting events, celebrations and urban vitality through special custom luminaries able to dynamically change colours (mainly the blue and magenta hues) in order to emphasize specific elements of the landscape, the green areas but also the human-sized areas (small paths and benches). Both functional and artistic lighting along with darkness were juxtaposed to create mystery to motivate further explorations: these lighting systems could create connections, comunicate, and remain in the memory as a sensation.

Finally, environmental qualities were associated to the use of high efficacy lighting fixtures and to the programming of flexible and transformable lighting performances, adapted to different urban social activities in different moment of the night and the seasons. This could decrease both lighting pollution and energy consumption when the area is not used.

7.4.2 The Holistic Approach Toward Social Oriented Lighting Design

The experience allowed the exploration of the holistic approach necessary to conduct complex urban lighting design with a social and human-oriented scope. Starting from the very preliminary stages of the project (analysis and concept development), three factors emerged as important when setting a successful social-oriented lighting design:

- the relation and appropriateness to the existing context. The ever-changing urban context should be studied to educate and instruct the design process and to adapt organically and resiliently to the environmental constraints of the site that change over time;
- the focus on the human dimension and the quality of human experience. The lighting performances have a strong relationship with the individuals providing connection with the space and surrounding environment through personalized and engaging experiences;
- the inclusion of people, either passive or active, in the design process. The study of people's activities, pattern of use, evaluation and appraisal of the environment and also of the urban lighting (existent before installation and the new one, after

installation) to understand how people live the space after dark and how they perceive the different luminous atmospheres.

The inclusion of these factors in the lighting design approach of the cities ensures that a generalist, engineering and uniformity driven conception is abandoned to embrace a more specific, finer, tailored and curated design of urban lightscapes. In this, the design process is transformed with new tools, strategies and methods.

7.4.3 Correlating with the Context: Diversity and Specificity

Exploring, analysing and understanding the context is fundamental to effective place-making. This practise is fundamental to move away from the implementation of uniform, one-size-fits-all designs based on compliant and universal norms toward the development of distinctive, bespoke and carefully curated night-time characters. Micro-scale, locally-responsive and contextual-based projects should explore and exploit all the different factors, features and conditions that make a space unique: local requirements, spatial qualities, historical contents, functional and environmental factors. Preliminary contextual studies could provide a deeper critical reflection of the local diversity and specificity, exploring information about public and private, used and underused spaces to incorporate this overlaid spatial consideration in the design strategies and concepts.

7.4.4 Focusing on Users: Quality and Experience

Along with the context, the user dimension should be explored, analysed and integrated when designing lighting for public spaces, in the early stages of projects. Human and social-centric lighting concepts should be based on a deep exploration of the heterogeneous needs and requirements of end-users, observing the activities and the multiple patterns of use happening in the nocturnal urban space. Observations and explorations serve to grasp how people use the cities and how they aspire to use it in the future. The people-centric approach offers the opportunity to apply social urban lighting strategies shaped to the specific beat of the city. This practice could define social benefits for different user groups and enhance positive behaviours, as well as decrease some difficulties and obstacles encountered nowadays in applying lighting technologies and solutions (e.g. the reluctance of elected officials for lowering lighting levels or reducing the operating time; the unmotivated negative impressions of citizens in relation to new lighting technologies). People could be involved also in the post-evaluation phase, to measure and assess the acceptability and impact provided by the implemented social-oriented lighting scenarios and performances.

In this regards, the "*Urban Social Lighting Design Radar*" could be also used as a post-evaluative tool, to inform the project by including the participation of different stakeholders and to assess the impact of the project (just simulated or realized after completion). Based on post-lighting observations (in situ if the project has been realized and based on photo-renders if it is not already installed), this procedure could be useful to critically review the project, capturing feedback either for a revision/integration/adaptation of the current project or for the development of future lighting schemes.

7.4.5 Transforming the Design Process: Integration and Collaboration

The interrelation between different competences and disciplines such as lighting design, architectural, urban and landscape planning, social sciences, sociology, ethnography, environmental psychology and many other is fundamental to converge toward the full potential of social-oriented lighting solutions. It is important to connect different expertise and integrate different points of view toward the same understanding. To envision scenarios that push the limits of new technologies and to answer the user and city needs, lighting designers should be prepared for a hybrid role, collaborating with other disciplines and facilitating the dialogue between different stakeholders. From this point of view, people from different disciplines and professions, with different methodologies, assumptions and talking different languages become equal stakeholders, together sharing the same vision and contributing to achieve a common goal: social oriented lighting projects. Lighting designers are in need of hybrid skills and tools to manage the social-oriented lighting experience as a whole process: exploring and designing actively the way in which lighting impacts human experience and transforms the urban space and vice-versa. In these hybridization of competences, lighting designers need to develop both soft and content-specific skills. About the latter ones, they need to develop a deeper understanding about interactive, responsive and adaptive technology based solutions, to achieve new luminous performances with augmented functionalities.

On the other side they also need to collaborate with social scientists to encourage a new democratic lighting design with citizens involved in the imagination and creation of their own lighting environments. This practice of engaging citizens and all the possible stakeholders beforehand or at the end of the design could be positive in fostering a critical attitude and a more aware consciousness of people about public spaces and lighting services. The social participation in the first phase of the lighting design is important to build a strong sense of belonging in the participatory definition of the lighting atmosphere between community members, neighbourhood associations, non-governmental organizations and other entities. The social inclusion during the evaluation of the project aims to gather the sense of lighting project in terms of achievements and benefits for people.

References

Briers G (2005) Lighting their fires through experiential learning. Agric Educ Mag 78(3):4–5

Cross N (2010) Designerly ways of knowing. Springer, London

Davoudian N (2019) Urban lighting for people. Evidence-based lighting design for the built environment, 1st edn. RIBA Publishing, London

Maile Petty M (2007) Light and the urban nightscape. Retrieved 17 May 2020, from https://www.archlighting.com/urban-design/light-and-the-urban-nightscape.aspx

Nelson H, Stolterman E (2003) The design way: intentional change in an unpredictable world: foundations and fundamentals of design competence. Educational Technology Publications, Englewood Cliffs

Seuntiens P, Vogels I (2008) Atmosphere creation: atmosphere and light characteristics. Philips Research, Eindhoven

The manufacturer's authorised representative in the EU is Springer Nature Customer Service Centre GmbH, Europaplatz 3, 69115 Heidelberg, Germany. If you have any concerns regarding our products, please contact ProductSafety@springernature.com

Printed and bound by CPI Group (UK) Ltd, Croydon, CR0 4YY
30/09/2026
02225468-0019